Aermacchi

Harley-Davidson Motorcycles

REDLINE BOOKS

Acknowledgements

This book, covering the Italian Aermacchi Harley-Davidson marque, is the third in the Redline Enthusiast Series and charts the history of the Varese-based company, which produced some truly magnificent motorcycles.

From the immediate post Second-World War era until well into the 1970s Aermacchi, together with its business partner Harley-Davidson from 1960, built a range of motorcycles which were admired world-wide. This was largely due to the marque's support of motorcycle sport, originally in trials, motocross and record breaking, but from the late 1950s, road racing.

Today, Aermacchi is very much part of the classic scene and is viewed by enthusiasts in a very special way thanks to its hard core of enthusiastic owners. Several of these have helped with information, and photographs; Pamela, Barry and Ian Lawton, Brian Clark, Terry Grotefeld, Dave Degens, Chris and Paul Bladon, Peter Fletcher, Mike Attrill, Bill Swallow, Andrew French, the Dutch-based Aermacchi Club and a special thanks to Dick Linton.

Mick Walker, Wisbech, Cambridgeshire

'It's a privilege to have access to the Mortons Motorcycle Media Archive, which I consider to be the finest in the world and a national treasure.'

MICK WALKER.

There is something very special about photographic images, the older they are, the more fasinated we become. Photographs are social and historical evidence which can give us a unique insight into the past..

We at Mortons Motorcycle Media, leading publishers of classic and vintage titles, own one of the largest photographic archives of its kind in the world, millions of motorcycle and related images, many of which have never seen the light of day since they were filed away almost 100 years ago.

Perhaps the greatest treasure is our collection of glass plates - almost two tons of them! They represent a largely unseen insight into our motorcycling heritage from the turn of the century onwards. These plates capture with tremendous clarity an era when the pace of life was much slower and traffic jams were unheard of.

We are delighted to be associated with Mick Walker in the production of this book and hope you enjoy the images from our unique archive.

Terry Clark, **Managing Director, Mortons Media Group Ltd.**

MORTONS
Media Group Ltd

ISBN 9780954435769

Brooklands Books

BROOKLANDS BOOKS LTD.
P.O. BOX 904, AMERSHAM,
BUCKS. HP6 9JA. UK

AERRB 1W6/3091

www.brooklandsbooks.com

Contents

REDLINE BOOKS

1 Before Bikes

Nieuport Macchi was founded in 1912 by Giulio Macchi. During the First World War the company built several types of aircraft based on French Nieuport designs.

Like Agusta (MV) and Piaggio (Vespa), Macchi came to motorcycles from an aviation background. Varese, a small provincial town near the Swiss border, has been a centre for the Italian aeronautical industry since 1912, when the firm of Nieuport-Macchi, or to give it its full title, Società Anonina Nieuport-Macchi, was established here by Giulio Macchi. The first two designs manufactured in quantity were the Nieuport Monoplane and the Parasol, a Macchi-designed monoplane using Nieuport construction methods. It is worth mentioning Nieuport was a leading French aviation company of the era, Edouard Nieuport's Bébé being noted for both speed and agility during the early days of aviation. Nieuport designs were some of the most successful in the formative years of war in the air. Fifty-six of the classic monoplanes were built during 1913-14, together with forty-two Parasol models. Both were powered by 80hp Le Rhône rotary engines, the monoplane Bébé capable of reaching nearly 70mph, and the Parasol slightly quicker. Both machines were used for reconnaissance and launched only a few hand-held light bombs. Both were straight copies of the original French designs. In 1915, Italy entered the war on the side of the allies, and Macchi began serious production of the famous Nieuport Sesquiplane, a biplane with the upper wing of considerably greater area than the lower. The first version saw 240 models built, equipping a fighter squadron at Aviano. Initially relying on small arms carried by the crew, they were later equipped with a light machine gun mounted either over the observer's cockpit or on the upper wing centre section for firing forward.

Next came the Nieuport 11, named the Bébé, a diminutive machine, which became the standard Italian fighter in 1917; and Macchi built 450. After this came the Nieuport 17, built in France during the First World War. But only 150 were built at Varese, before production was phased out in favour of the Hanriot HDI. From November 1916 until production ceased in 1919, a total of 831 Macchi HDIs were constructed.

But it was seaplanes for which the fledgling Macchi concern was best known during its early years. This came about in rather an unusual way. Austrian Lohner flying boats had proved very successful over the Adriatic. When one was forced down on the water near the Porto Corsini Seaplane base in May 1915, it was transported to Varese, where the Italian company produced a copy of the Austrian design. The result was the Macchi L1. And so came about the L series. The Mark 1 and 2 machines were essentially the Austrian design, but for the L3, Macchi used its own Nieuport influenced design skills carried out by Felice Buzio. As many as 360 L1, L2 and L3 machines had been manufactured by the end of 1918. The L3 continued in service to the Regia Marina (Italian Naval Air Service) until as late as 1926.

The very successful Macchi M5 single seat fighter seaplane, first flew in 1916.

The Fiat liquid-cooled V12 engine used to power the Macchi Schneider Trophy seaplanes.

But the real Macchi success story was the M5, a lightweight single-seat fighter flying boat, which first flew in 1916. This was one of the few machines which could rival the best land plane fighters. With a 170hp Fiat engine, maximum speed was 127mph. Including the M5 Mod (Modified), a total of 340 M5 series aircraft was built. Production of the later M7 continued through the 1920s. Other Macchi seaplanes followed, both military and civilian.

Perhaps the most famous of all were the Macchi seaplanes which took part in the prestigious Schneider Trophy contests. The M7 biplane of Giovanni de Briganti won the contest in 1921, averaging 117.8mph over the 10 lap, 230 mile course. The following year one of the same Macchi seaplanes finished runner-up, with a newly designed Macchi M17 third, both beaten by a British Supermarine Sea Lion. The next Macchi bid for honours came in the USA in 1925. There was a new machine, the M33, designed by the firm's promising young designer, Mario Castoldi. But against huge American competition Giovanni de Briganti could only finish third. Mussolini was now in power, and the Italian government put great store by events such as the Schneider Trophy. For 1926 Castoldi designed the new M39. Capable of almost 260mph, it was a serious contender and the first non-American to win the trophy in the States, with Maggiore Mario de Bernardi winning at an average speed of 256.5mph. Team mate Tenente Bacula came third in another M39. The reputation of Aeronautica Macchi (the company having dropped the reference to Nieuport in 1922) was growing in stature.

Then, on the 17 November 1926 de Bernardi shattered the world speed record over a 3 kilometre course, with a new speed of 416.6kph (258mph).

By 1927 another Macchi was ready, the M52. Although powerful, all three Macchis had to retire with engine problems. The British Supermarine S5 was victorious.

The next contest, at Spithead saw the new Castoldi-designed M67, with a powerful 1,400 hp Isotta Fraschini 18 cylinder engine. But disaster struck on the 22 August 1929. During testing, Capitano Motto crashed into Lake Garda and was killed. However, Macchi still competed in Britain, with Tommaso da Molin claiming runner-up spot behind the Supermarine S6 of Flight Lieut. Waghorn.

Then on 30th March 1928 de Bernardi established a new world speed record in an M52R at 512.78kph (318.63mph).

For the 1931 Schneider Trophy, Mario Castoldi designed the new MC72, but unfortunately it was not ready in time, so the British, with three victories in a row won the trophy outright. Nevertheless, development of the MC72 continued. This determination was to pay off. Early in 1933 the MC72 was finally ready, its engine producing an amazing 2,600 hp. On the 10th April 1933 Warrant Officer Francesco Agello attained 682.1kph (423.8mph). Then on 8th October that same year, Guglielmo Cassinelli averaged 629kph (391mph) over a 100 km (62.5 mile) course with the Fiat AS6 engine boosted to deliver 2,800bhp.

The MC72 broke the absolute world speed record in October 1934 at over 440mph.

First flown at the end of 1937, the Macchi MC200 Saetta (Arrow) was built in larger numbers than any other Italian monoplane fighter of the Second World War. It is recognised as being part of the best family of aircraft operated by the Regia Aeronautica during the conflict.

Macchi MC202 Folgore (Lightning) was powered by a licence-built version of the German Diamler-Benz liquid-cooled V12 engine and could reach almost 400mph.

But the crowning glory for Macchi, the MC72 and for designer Mario Castoldi, came on the 23rd October 1934, when, with the AS6 delivering 3,100bhp, Agello reached an amazing 709.2kph (440.7mph) over a 3 kilometre course in a MC72, registration number MM181, establishing an absolute world speed record which was to remain unbeaten by a landplane for the next five years.

This success put Macchi at the very forefront of aviation technology, but also established Mario Castoldi as one of the world's top designers. The big question was, could equal success be achieved with mass-produced civilian and military aircraft?

In many ways Mario Castoldi can be compared to Reginald Mitchell. Like Mitchell, Castoldi designed ultra fast seaplane racers. And like Mitchell, Castoldi created a low-wing monoplane fighter prototype in the mid 1930s. The big difference was that the British designer had the use of a powerful liquid-cooled inline engine (the Rolls Royce Merlin), whereas Castoldi was saddled with drag-inducing and less powerful radial power plants. Unfortunately none of the experimental high-powered Fiat engines from the Schneider racers evolved into one for normal production use. Thus, with its bulky A74RC38 radial with bulged rocker arm covers the Castoldi prototype of what was to emerge as the MC200 Saetta (arrow) could only reach 318mph, compared with 350 plus for the Mark 1 Spitfire. Not only this, but the British aircraft had 8 machine guns, and the MC200 had only a pair 12.7mm Breda SAFAT machine guns with 370 rounds per minute. Nonetheless, the MC200 did represent an advance over its biplane predecessors, with an all-metal construction and semi-monocoque fuselage structure, whilst its undercarriage retracted inwards to lie within the wing profile.

As later events were to demonstrate, the airframe was an excellent design, but was hampered by the non-Macchi power plant.

The famous lakeside Schiranna factory on the outskirts of Varese, home of the Schneider Trophy winning seaplanes and world speed record breaking MC72. The site with its massive hangar and slipway became the headquarters for Aermacchi Motorcycles, before becoming Aermacchi-Harley-Davidson in 1960. Today Cagiva motorcycles are built there.

This was finally redressed to some extent in 1940 when Castoldi obtained a German Daimler Benz DB601 engine, the same unit which powered the Messerschit Bf 109. Thus Macchi began a private venture to build a DB601 powered version of Castoldi's design. This was to emerge in 1941 as the MC202 Folgore (Lightning). With superb manoeuvrability, a top speed of 370mph and the robustness of the MC200, it was considered by many to have been the best Italian fighter aircraft of the Second World War. But too few could be built to make a real difference, due to production glitches of the Fiat DB601A engine, not Macchi's fault. Even so a total of 1105 MC202s had been delivered prior to the September 1943 armistice.

A development of the MC202, the MC205 Veltro (Greyhound) had first flown in prototype form on 19th April 1942. This was essentially a modified MC202 airframe with the more powerful DB605A engine, which put out 1,475 hp

Post Second World War the Aermacchi MB326 advanced jet trainer and ground attack aircraft was a huge commercial success, flown by several of the world's airforces for many years from the early 1960s.

Unable to manufacture its traditional aviation products, Aermacchi's first postwar offering was the MBI three-wheel truck. Powered by a 750cc twin-cylinder four-stroke engine, it proved an immediate success and remained in production for over four decades with very little change, except for an increase to 973cc and a switch to a diesel engine.

against 1175 hp of the DB601A. With a maximum speed of 399mph, the MC205 could have been a decisive weapon, but production did not match demand, only 262 aircraft were produced, and mostly after September 1943.

The final development in the MC (Mario Castoldi) series was the MC205N Orine (Orion). This was an even more advanced design, incorporating considerable refinement, differing principally in type and arrangement of the armament; there being three 20mm cannon and two 12.7mm machine guns. Other projects were the MC206 and MC207, but neither reached the production stage before the war ended.

Because of its involvement with the supply of military goods, notably aircraft, Aermacchi's various production facilities in and around Varese were a prime target for allied bombers during the conflict. Then, in the immediate post-war days they were the subject to strict controls, with aviation production banned for several months. Subsequently, the company was able to resume its position as the major Italian aircraft manufacturer, with licence-built aircraft including the British de Havilland Vampire jet fighter/ trainer and Dutch Fokker S-11 primary trainer (designated M416). In addition Castoldi had been replaced as chief designer by Ermanno Bazzocchi. But this return to aviation did not come quickly enough to keep the workforce busy during the late 1940s.

To rectify this situation, the 'Macchi' management was forced to consider other outlets for its undoubted engineering skills. The result was a decision to design and manufacture a three wheel commercial vehicle, followed by (as described in the next chapter) powered two-wheelers.

The three-wheeler was powered by a 750cc four-stroke twin cylinder engine and

Gruppo motore cambio

later by a 973cc diesel unit. Known as the Macchitre MBI, this proved an outstanding commercial success and was to remain in constant production until well into the 1980s, albeit by that time no longer produced by the Varese Aermacchi plant. The MBI series should not in any way be confused with the Motocarro, a motorcycle-based lightweight goods vehicle produced during the 1950s, described in Chapter Two.

By the late 1940s, Aermacchi was not only able to consider its future with far more confidence, but its diversity was the envy of many competitors, with its truck operation, a new range of motorcycle underway, a marine division and a return to its core business, aviation.

Interestingly, during the 1950s Aermacchi became closely involved with the American Lockheed Corporation and it was this association, which was to play a role in the motorcycle division's decision to link up with another American company, Harley-Davidson, in 1960.

2 Early Motorcycles

Aermacchi's first ever two-wheeler was this prototype electric-powered scooter, 1943.

Little known double overhead camshaft single cylinder racer of 1951 intended as a challenger to the likes of MV Agusta and FB Mondial at Grand Prix level. However, it didn't prove fast enough.

In 1943 Aermacchi took its first step toward motorcycle manufacture with the production of a prototype electric motorcycle.

Anticipating the post-war motorcycle boom and intending to have a stake in it, Aermacchi bosses looked for a competent designer who could come up with an attractive lightweight. Their choice was Ing. Lino Tonti, who had been at Benelli and had worked during the war in the aviation field on aero-engines.

Tonti's first design in 1950, for his new employers, was decidedly unorthodox. It was a lightweight with a 123cc (52 x 58mm) two-stroke engine of full unit construction, 3-speed gearbox, and a horizontal cylinder. Maximum power was 5bhp at 4500rpm, giving a top speed of 47mph. The frame was of the open type in which the engine pivoted in unit with the rear suspension. The luggage or tool container, shaped in the fashion of a conventional fuel tank, featured a hinged mounting at the steering column, allowing the rider to grip something with his knees. The real tank was to be found under the saddle. This machine which looked like a scooter with motorcycle-size 17in wheels, was named the 125 and was very much in the same vein as the contemporary Moto Guzzi Galletto.

Although never as popular as the larger-engined Guzzi device, the Aermacchi sold well enough to encourage further development, which also included forays into the sporting world with specially prepared versions in the 1951 and 1952 ISDT. By 1953, the model range had expanded to include three versions of the 125 - the basic U model, deluxe N, and much more attractive M sportster. There was also a 246cc '250'. Essentially, this was two 125s coupled together, mounted in a 'U' shaped tubular frame and with a 4-speed gearbox (the smaller models retained only 3-speeds).

Using the existing 123cc cylinders (and pistons), the 250 engine produced 10bhp. Other details of the machine's specification included; 15 inch wheels, single sided brakes, a 'banana' shaped tubular frame, pivoting front and rear suspension, spring saddle and separate pillion seat, aluminium wheel rims, 65mph maximum speed and a dry weight of 97kg (214lb). Unfortunately, poor sales resulted in production of the 250 Aermacchi twin being axed during 1954.

Left, one of the earliest production models, the 125N, (a 1953 machine is shown). Above the 125M, a strange mix of scooter and motorcycle components with a two-stroke single cylinder engine.Together with the 125U it laid the foundations for the future.

The 125U with trailing link front forks and cantilever rear suspension. An advanced, but unusual design.

1954 250 Twin

Engine:	Air-cooled piston-port horizontal twin cylinder two-stroke, alloy head, cast iron barrel, unit construction.
Bore:	52mm
Stroke:	58mm
Displacement:	246 cc
Comp ratio:	6:8:1
Lubrication:	Petroil mix
Ignition:	6 volt flywheel magneto, 2 coils, 8ah battery
Carburettor:	Dell'Orto
Primary drive:	Helical gears
Final drive:	Chain
Gearbox:	4-speed, foot change
Frame:	Tubular, U shape
Front suspension:	Pivoting fork
Rear suspension:	Pivoting fork with friction damping
Front brake:	Drum, single-sided hub
Rear brake:	Drum, single-sided hub
Tyres:	3.50 x 15 front and rear

General Specifications

Wheel base:	37.4 inches (950mm)
Fuel capacity:	3.63 gal (16.5ltr)
Dry weight:	214lb (97kg)
Maximum power:	10bhp@5,000rpm
Top speed:	65mph (105kph)

This interesting, 250 twin was introduced in 1953. It employed two of the existing 123cc cylinders to make 246cc. With bore and stroke dimensions of 52 x 58mm, the engine produced 10bhp. Other details included 15in wheels, 4-speed gearbox and dry weight of 97kg (214lb), maximum speed was 65mph (105kph). Poor sales saw it axed before the end of 1954.

The ultimate development of the famous U-frame Aermacchi range of two-stroke lightweight motorcycles was the Zeffiro. Initially, this machine which arrived in the mid 1950s, was very similar to what had gone before, a cross between a scooter (weather protection) and motorcycle (wheel size). Nonetheless the Zeffiro was an attractive, modern-looking machine, whereas several of the earlier Macchi designs had been quite basic, even unattractive, though they might well have been functional.

Certainly, the engine unit, with its horizontal cylinder and piston port induction, plus the 3-speed twist grip operated gearbox hadn't changed much. However, the styling certainly had. This was helped greatly not only by much smoother lines, but also the introduction of telescopic front forks. The first examples of the Zeffiro had single sided wheel hubs, however these were eventually replaced by full width highly polished aluminium components. In both cases the hubs were laced to 17 inch chrome steel rims.

Both mudguards were deeply valanced, whilst the headlamp was neatly incorporated into the enclosed handlebars, which also encased the speedometer (very much as seen on the best selling Lambretta and Vespa scooter several years later). This meant that the light beam moved with the rider when the bars were moved. At first the telescopic forks featured rubber gaiters at

This 350cc prototype appeared in 1955 and owed its styling to the American automobile of the period. Like the production 250, it featured a twin cylinder two-stroke motor with horizontal cylinders. Only one example was constructed.

The Zeffiro, the final derivative of the combined scooter/motorcycle Aermacchi 125cc family, mid/late 1950s.

their base. However, later the forks were extended below the wheel spindle line and lost the gaiters. A horn was incorporated centrally into the steering stem below the headlamp and above where the front forks spread out either side of the mudguard/wheel. Scooter-type leg shields extended rearwards at this base to provide excellent rider protection, at least for the legs, feet and lower body. The frame was of the 'step-thru' type (as on a scooter) and thus there was no conventional tank to press the knees against. Instead the tank was found under the rider's seat (a single, sprung type as one would have found on a Lambretta Li scooter). A small parcel carrier was incorporated behind the seat.

A larger displacement (150) version of the Zeffiro was also offered. This proved, like the 125, to be a popular machine. Both continued to be offered into the newly 1960s (the only two-strokes to survive at this time, if one discounts the released Brezza Scooter). The 150 Zeffiro had an exact engine capacity of 147.9cc (57 x 58mm) and could reach 90 kilometres an hour (57mph).

Like its smaller brother, the 150 version had the now familiar Aermacchi horizontal cylinder layout and, as on the 125 version, a 3-speed twist grip operated speed box.

1959 Zeffiro 125/150

150 in brackets where different

Engine:	52mm (57mm)
Stroke:	58mm
Displacement:	123 cc (147.93cc)
Comp ratio:	6:1 (7:1)
Lubrication:	Petroil mix
Ignition:	6 volt, flywheel magneto
Carburettor:	Dell'Orto
Primary drive:	Helical gears
Final drive:	Chain
Gearbox:	3-speed
Frame:	Step-thru, all steel construction
Front suspension:	Telescopic fork
Rear suspension:	Hydraulically damped shock absorbers
Front brake:	Full width, alloy drum, SLS
Rear brake:	Full width, alloy, drum, SLS
Tyres:	Front 2.00 x 17, rear 2.50 x 1

General Specifications

Fuel capacity:	1.87 gal (8.5ltr)
Dry weight:	172lb, 78kg (176.5lb, 80kg)
Maximum power:	6.5bhp@5,500rpm (8.5bhp@5,500rpm)
Top speed:	125:49mph, 76kph (57mph, 90kph)

But there were notable differences between the two versions; the bigger model had a redesigned sturdier frame, with revised rear suspension and a dual seat assembly. Like many other Italian makers, Aermacchi also offered its own Motocarro (light truck). This employed the front end of the 150 Zeffiro, including not only the engine assembly but also some of the cycle parts such as bare front forks, front mudguard and leg shield assemblies, to mate up with an entirely new rear half, half the truck section. This had a maximum loading capacity of 350kg. Unlike the conventional two-wheel Zeffiro, the three-wheel commercial vehicle had a much more spartan headlamp/ handlebar section. Also the larger battery was left exposed, whilst there was now a foot-operated heel and toe gear change mechanism. But an additional gear ratio had been added, making four in all. The wheels were of an entirely new disc type, with 3.50 x 15 tyres.
As an the standard Zeffiro, the engine was fan cooled and a kick starter was used to get it under way. There was a dry, multi-plate car type clutch, whilst the rear wheels were driven by a shaft with a flexible (universal joint) coupling.

Another Aermacchi product of the mid-late 1950s was the Cavaliere. The 3-wheel light truck was powered by a 148cc (57 x 58mm) single cyilnder two-stroke engine.

A 125cc two-stroke built in 1957 and very much in the mould of the Chimera (see chapter 3).

Rear suspension was made up of leaf springs and a coupling bar, whilst there were hydraulic brakes on the rear wheels, with the conventional mechanically operated drum on the front wheel.

The remainder of the specification for the 3-wheel truck, the Cavaliere, was as follows;

Ignition	flywheel magneto 6 volt, 28 watt
Lighting	battery with rectifier
Fuel	petrol with 6% oil mixture
Dry weight	190kg
Fuel capacity	6 ltr
Performance	6.5bhp at 5,000rpm
Max speed (with full load)	55kph
Maximum height	9,150mm
Height of loading platform	550mm
Maximum length	1,330mm
Wheel base	1,677mm

In 1955, Tonti designed a pair of record-breakers, making full use of the wind tunnel and other resources of the company's aviation division. Powered by double overhead cam engines of 49 and 75cc, the larger machine was one of the first examples of a 'flying cigar', being extremely low to the ground and some 3 metres long with a special hand-beaten aluminium streamlined shell. Positioned behind the rider, the 75cc motor had bore and stroke dimensions of 44.5 x 43mm and running on a 10:1 compression ratio, gave 9bhp at 11,000rpm.

The Belgian Market

In Belgium the name Aermacchi is inseparable from that of Defco, Liege. From 1959 Defco was the importer of Mi-Val, Capriolo, Beta, Alpino and Freccia Azzura.

While on holiday in Italy, M. Noseent, a director of Defco, saw a scooter in the window of a motorcycle dealer in San Remo. Having identified it as an Aermacchi type 125 M with hinged tank, M. Nossent recognized the advantage of this scooter which could be imported in parts, thus avoiding the high import duties of the time. In Spring 1953 he went to Aermacchi's head office in Milan to arrange a distributorship with them. He was successful and in 1954 Defco exhibited an Aermacchi on their stand at the Brussels Motorcycle Salon.

In the following years, Defco expanded the range of Aermacchi models. In 1955 came the 125 U and 125 N and in 1956 the Ghilbli 125 and Zeffiro 125 were added, the latter also available in a 150 cc version. These Aermacchi scooters were particularly popular among the clergy in Belgium. By 1959 the Chimera, Ala Bianca, Ala Azzura, Ala Rossa and Ala Verde were added to the range.

Defco continued as the Belgian Aermacchi importer until 1973. In that year, control of Aermacchi became 100% American and M. Nossent saw the end of his distributorship in sight because his corporate strategy and spare parts stocking policy did not comply with Harley-Davidson's plans. During 20 years of their agency, Defco had imported about 300 Aermacchis, of which twenty were Ala d'Oro.

The first Ala d'Oro sold in Belgium was delivered in 1969 to Christian Michaelis. Until then, the Belgian 250cc racing class was dominated by Vervroegen on his 203cc works Motobi, followed by Hanset (Maico) and Albert (Adler). Michaelis took first place at Moeskroen, heralding a new era.

Another historic event in 1960 was the debut of Aermacchi in an international GP at Spa-Francorchamps (Pagani No. 40). The Italian took fifth place on this practically unknown machine behind a bunch of works MVs and Mike Hailwood on a Ducati, to launch the Aermacchi onto the 'Continental Circus'. As the season progressed, Michaelis got better results, 2nd behind Terlman at Erteveld and third at Hesbay.

Although Piet Vervroegen was the best rider in the 250 class, he knew that the Aermacchi onslaught could not be halted. He therefore ordered three new Aermacchis from Defco at the end of the 1960 season. These were the Ala d'Oro S type, with two long strokes (66 x 72mm) of 246cc and an experimental 'square' (72 x 72mm) 293cc machine.

The 1961 Pegase racing team consisted of Vervroegen, Hemptinne and Dellelio. They fought for the highest honours in those days when it was quite common to ride a 250cc machine in the 350cc class and not without success. The 350 class consisted mainly of Manx Nortons and 7R AJS often at a disadvantage against the lighter Italian 250s.

The Pegase Equipe entered both the 293 and the 246 Aermacchi in the 350 class; using the lightweight machine to keep the competition at bay. The 250 withdrew from the race a few laps from the end, allowing the 'Junior' Ala d'Oro to go on to victory in nearly all 350 events of 1961. First Vervroegen and then Hemptinne would win. Michaelis was dogged by gearbox problems throughout the year and a 3rd place at Moeskroen was his best performance.

The race department at Varese cured the gearbox

problems as they occurred and at the end of the season Vervroegen won the Belgian 250cc Championship just as in 1959 and 1960.

From 1962 onwards several new Aermacchi riders appeared in the Belgian championships, including Orinel and Petroons; the latter on a race-kitted Ala Verde. In the national race mainly the invited foreign riders set the pace. At Moeskroen (6.5.62) Joop Vogelzang won on an Aermacchi from Vervroegen who had plug problems. At Chimay, works riders Pagani and Milani came to teach the Belgian riders a lesson. Pagani won easily on a factory machine at 159kph.

Highlights of that year included sixth place by Vervroegen in the Belgian GP at Spa. He was second at St Truiden behind the flying Vogelzang with Dellelio fourth and Michaelis eleventh.

At the opening of 1963, the oil giant Mobil withdrew its racing support and because of this Vervroegen retired from racing. He sold his Aermacchi to Van Oudekerke who was sponsored by Demeurichy. Florent Verdoodt and Erik Goegenbeur bought new Ala d'Oro 250 S machines. Michaelis acquired a new short- stroke engine from Varese for Bf 24,000 after blowing up his faithful long-stroke on the ultra-fast Spa-Francorchamps circuit. At that time, it was necessary to lap at more than 150kph (94mph) in order to qualify for the Belgian GP. Since Michaelis had averaged only 149.920 he thought he had failed and turned up on Sunday as a spectator, only to be informed by Toussaint that his entry had been allowed. At this point, his long-stroke engine was in pieces, having been dismantled the previous evening to investigate problems. In order to race, he had to reassemble the engine. Because he was in such a hurry, Michaelis accidentally under-geared the bike, which allowed the engine to over-rev and melt the piston so that virtually nothing remained of it. That was the end of the long-stroke.

The most important results in 1963 were a first at Moeskroen by Michaelis, a first by Verdoodt at St Truiden, not forgetting the fantastic performance at Beveren Waas by Visenzi. Michaelis was now the most regular of the 250 riders.

Early in 1964 Verdoodt was seriously injured in a car accident when returning from Varese, but fortunately recovered in time to ride the 350 series at Moeskroen. The 250 series was not held in 1964. However, the season ended with Toussaint (Morini) as 250 National Champion ahead of Michaelis (Aermacchi) while Belgians Raymond Bogeart and Serge Dupont also rode new Ala d'Oros that year.

At Moeskroen in 1965 Visenzi again had a fantastic race in the 250 class, taking 2nd place on his 250 in the 500 race behind John Smith on a Norton. Moeskroen in that year was a real festival for the Aermacchis, as they took the first eight places; Visenzi, Barbaroux, Bogaert, Blanchard, Leenheer, Goegenbeur, Van Oudekerke and Michaelis. By this time it was plain that the Aermacchi was down on speed on the fast circuits such as Mettet and Spa when pitted against Honda, Suzuki, Yamaha and Bultaco. Raymond Bogaert was fourteenth at Spa.

On the short twisting circuits Aermacchis were still able to hold their own, for example Michaelis and Van Oudekerke took first and third Neufchateau. At Beveren-Waas, Van Dongen (Bultaco) finished first ahead of De la Porte and Bogaert (Aermacchi). The 1965 National Championship was won by Bogaert with Michaelis second.

The opening race at Moeskroen in 1966 was won by Anderson on a Husqvarna with Michaelis third and Verdoodt sixth. Bogaert, the Belgian National 250cc Champion, crashed at Mettet at 180kph, fortunately without serious consequences. Not so his Aermacchi, which needed a rather comprehensive repair job.

1966 was very successful for Michaelis, who ended the season as National 250cc Champion. At the beginning of the year he had bought a new dry-clutch Aermacchi with the redesigned frame. Following conrod failure at Zolder the engine was repaired at Varese and some new exotic details added. The results were quite promising, third place at Moeskroen, a second at Bois de Villers and third at Natoye.

From 1967 onwards the Aermacchi was not fast enough at national level and could no longer make the grade. The Ala d'Oro was simply too slow when pitted against two-strokes such as Bultaco and Yamaha. Because of this and because of the disappearance of most of the local short circuits, many riders sold their Aermacchis. Some of the by now obsolete Italian four-stroke singles were used in the national Hill Climbs which were popular in the late sixties and early seventies.

Nevertheless the faithful continued to race their Aermacchis the familiar names of Michaelis and Verdoodt still featuring in the midfield results.

My thanks to Christian Michaelis, Van Eycken and M. Nossent for this insight into the Aermacchi scene in Belgium.

Aermacchi broke a number of speed records in 1955. These tiny streamliners piloted by Massimo Pasolini and designed by Ing Lino Tonti, were built with 48 and 74cc overhead camshaft engines.

The camshafts were driven by chain, and the gearbox had four ratios. Primary transmission was by gear, and final drive was taken care of by a chain.

The frame was a central spine type, the front suspension featuring long moveable levers with twin shocks. There was a conventional swinging arm at the rear. A mixture of pump petrol and alcohol was used for fuel, whilst tyres were 2.00 x 18 on both wheels. The smaller machine gave 7bhp at 12,000rpm from its over-square 40 x 39mm bore and stroke dimensions, and was quite different in two important areas; it featured a horizontal cylinder, and didn't have the same frame or fully-enclosed streamlining of the larger engined machine. Instead a conventional frame with separate front and rear fairings was employed.

The actual record attempts were made on 4th April 1956, over a section of the Milan-Varese autostrada. Neither weather nor road conditions were ideal, but the 75 version, piloted by Massimo Pasolini (father of future racing star, Renzo), reached a speed of 100.2mph over the flying mile, and 108.8mph over the flying kilometre; breaking by a considerable margin the records set up only a few months before by Germany's Adolf Baumm with his NSU 'flying hammock'. With the 50, the standing mile record was broken with a marked increase, at 51.25mph.

Soon after these successes, Ing Tonti left to join the rival FB Mondial factory. His place as technical director at Aermacchi was filled by Ing Alfredo Bianchi, previously with Parilla and the Alfa Romeo car company. Whilst at Parilla, Bianchi had assisted in the design of the various single and double overhead cam sports and racing single-cylinder machines with which Parilla had garnered so much success in the late 1940s and early 1950s. In addition, he had also manufactured his own Astoria engines and complete machines, the single-ohc and two-stroke power units enjoying a fair degree of popularity as proprietary fitments for use by smaller manufacturers.

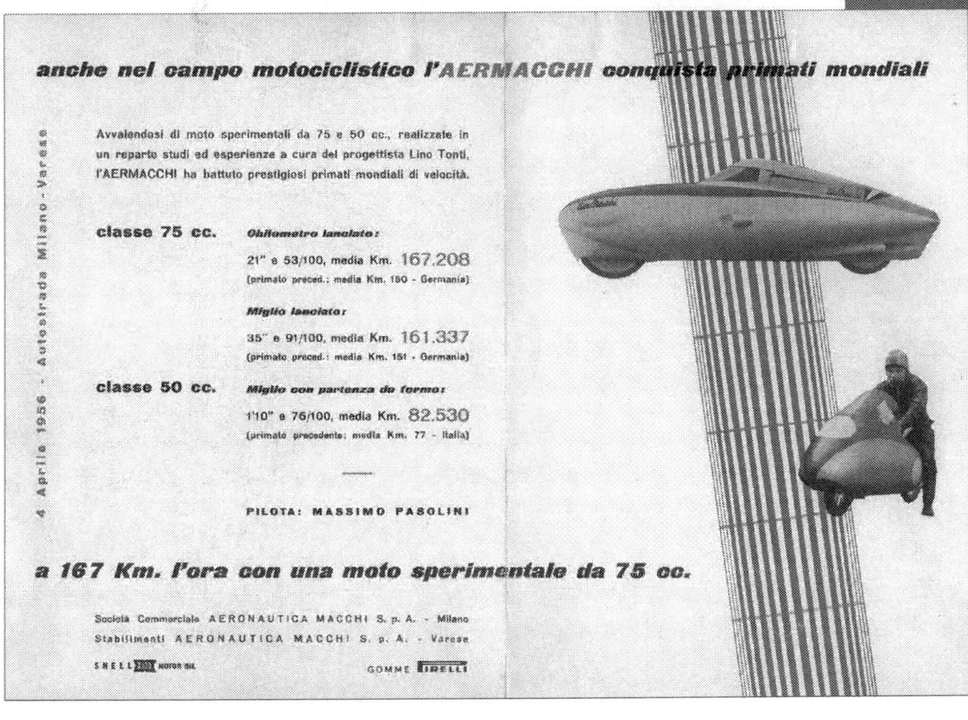

The record attempts were made on 4th April 1955, on the Milan-Varese autostrada. Piolted by Massimo Pasolini (father of Renzo), the larger 74cc engined streamliner broke the class records for both the flying kilometre and flying mile by a considerable margin.

3 Chimera

As mentioned in the previous chapter, Lino Tonti was replaced as Aermacchi motorcycle design chief by Ing. Alfredo Bianchi in early 1956.

Bianchi's first task on joining the Varese concern was to evolve a production machine using as a basis a sketch of an ideal motorcycle which had been executed by Count Mario Revelli, a well-known car stylist who had himself been a racing motorcyclist, and who had actually won the 1925 Italian Grand Prix at Monza on a CR bike of his own manufacture.

When launched at the opening of the Milan Show on Saturday 1st December 1956, the debut of the new Aermacchi, the Chimera (Dream), created a sensation. The British journal, The Motor Cycle, reported on the Chimera in its 6th December issue. 'It is an outstanding example of original design and engineering craftsmanship, the like of which has rarely been equalled in the motorcycle sphere. Words and photographs barely serve to get across the unique nature of the machine. It has to be seen to be appreciated to the full; to be admired as it deserves to be admired. It is a design in which pressure die-casting blends homogenously into sheet-steel pressing. Line follows line in sleek, unbroken, harmonious sequence. From nose to tail the machine presents a picture of discreetly refined engineering elegance and more than that - of the highest degree of craftsmanship.'

In spite of its ultra-modern appearance, the Chimera's stylish overcoat hid a fairly orthodox engineering package. The ohv power unit was slung below a tubular spine frame, at the rear of which was a fabricated structure carrying the rear-fork pivot. Sitting on the frame was the main body section, an elegant unit built up from steel pressings, and comprising the rear mudguard and fuel tank. The nose of the unusually long seat blended into the tank, and the elongated kneegrips assisted the styling.

Forming a duct round the near-horizontal cylinder was a light-alloy die-casting, which merged into the mid-section fairings. Instead of being added panels, the fairings were in fact the engine unit's side-castings, suitably extended and profiled. Bridging the space between the engine enclosure and the fuel tank were two more light-alloy die-castings which met at the front where they embraced the horn. Attached to the rear of the engine side covers, and supported on struts projecting rearward from the fabricated frame structure, were matching pressed-steel fairings. The one on the nearside shielded the otherwise exposed rear drive-chain, and both served to carry the lines of the engine enclosure through to the rear wheel. All the bodywork, with the obvious exception of the engine side covers, was readily detachable for accessibility.

Rear suspension was controlled by a single hydraulically-damped unit, lying above the rear end of the frame spine and operated by an arm welded to the fork bridge. A conventional set of telescopic forks carried the front wheel, and both wheels had 17inch chrome-plated steel rims and light alloy, full-width brake hubs.

The headlamp unit and speedometer were encased in a light alloy casting which formed the upper yoke of the fork. Behind the speedometer was fitted a second, quickly-detachable casting enclosing the steering head, control cables and most of the handlebar, the twin castings combining to provide a remarkably clean appearance.

At a time when the vast majority of Italian designers regarded overhead camshafts as absolutely imperative even for touring models, Ing Bianchi had given the Chimera a pushrod motor. This was primarily in the interest of simpler maintenance; but he was also of the opinion that, even for sportsters, there was no real need for the ohc layout. And, as events were to prove, this theory was not far off the mark.

Initially, the Chimera was only offered in one engine size, this being 172.4cc (60 x 61mm). With a compression ratio of 7:1, maximum power was a respectable 13bhp. Other details included a roller bearing big-end, light alloy cylinder head with enclosed valve gear, crankshaft-mounted generator, coil ignition with auto-advance, wet sump lubrication, 4-speed gearbox, multi-plate clutch and helical gear primary drive.

When British journalist John Thorpe visited Italy for the 1956 Milan Show, he also travelled to Aermacchi in Varese, and he described it as the 'second largest Italian Aircraft concern who, have found motorcycles a sound commercial proposition'. John Thorpe's report of his Aermacchi visit was published in 'MotorCycling' dated 27 December 1956. He described the 'Macchi' empire thus;

'There are four separate plants in the Aermacchi set up. The chief is the aircraft and motorcycle factory in Varese itself. Also in the town is an assembly plant for the three wheeled delivery vehicles and motorcycles. Then, on the outskirts of Varese, at Schiranna, is the old seaplane factory, with its slipways into the lake, and its left hangar. This now provides storage space for motorcycles.'

These were the two facilities Thorpe visited in 1956, the others being an aircraft assembly and airfield at Venegono, and a shipbuilding arm, the Cantiere Navale di Apuania.

'One factory is generally much like another, but there is a distinct difference between Aermacchi and most of the other motorcycle factories which I have

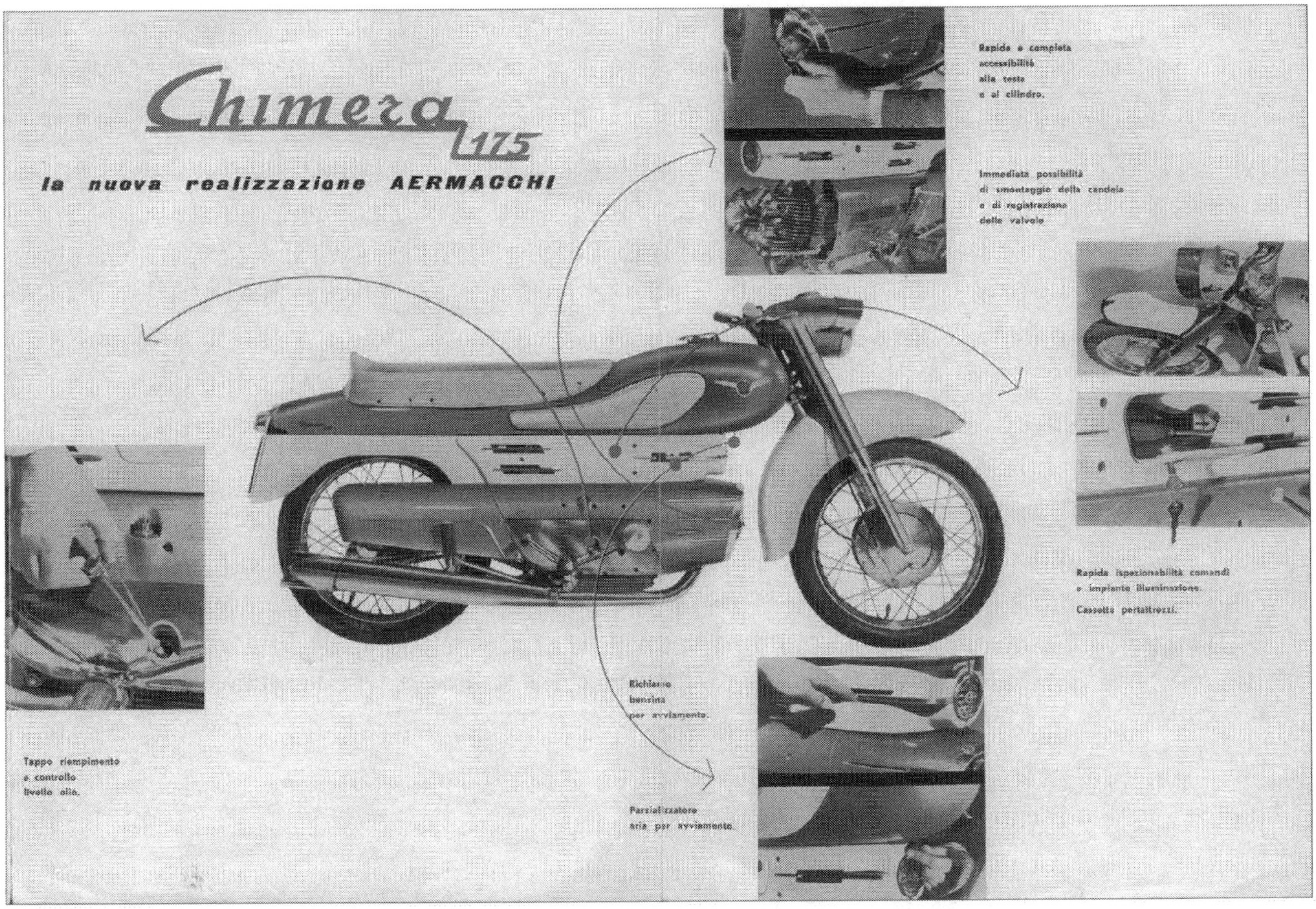

visited, in that one is forever tripping over pieces of aeroplane! In the huge, common machine room, for instance, motorcycle, delivery truck and aircraft parts were all being worked upon at once. The same machines are employed. The same operators deal with motorcycle and aircraft components. The same control tests are applied to both. Since the aircraft side demands stringent tests, and strict adherence to precision standards, the benefit to Aermacchi customers on our more mundane level is at once obvious. A case in point is the fully equipped test shop, designed to carry out exhaustive experiments to ensure that the material which goes into the jets is 100 per cent right. This shop samples all material, which is bought, regardless of whether it is to be used for aircraft or motorcycles. Obviously, that ensures the right beginning, metals that are sound to start with.' There was also a particular mention of the newly launched Chimera. 'At Schiranna I found, tucked away in the experimental shop, a fascinating frame-breaking machine on which the Chimera's frame had been tortured for months before it was accepted for production. In theory, it is something like the suspension demonstrators which you see at Earls Court. The frame was anchored by steel straps to a bench, while huge cams pumped the suspension, and other little devices applied stress at such points as the steering head. It brought out initial weaknesses in the rear suspension springs, but I was assured that at no time had the frame itself failed.'

Strangely, even though Thorpe had actually visited Aermacchi (meeting with Dr Filipo Pedote), it was 'Motor Cycling's' rival, 'The MotorCycle', which published the first British test of the sensational newcomer. This appeared in the 11th July 1957 issue. The article began, 'It comes as no surprise to learn that the Aermacchi Chimera was designed by one of Italy's leading car stylists. The daring originality of line and construction are refreshingly free from conventional ideas. Though its styling is altogether too bold and novel for the model to be described as typical of contemporary Italian design, the flamboyance of line and colouring (salmon pink, cream, polished aluminium and chromium plating) is unmistakably Latin. Except for the unusually deep mudguards, the effect of the enclosure is more aesthetic than functional. That is to say, there is no attempt at built-in weather protection, though shrouding the engine does, of course, make for cleanliness in use.

'Two of the biggest problems facing the designer of an enclosed model concern engine cooling and accessibility. Both difficulties have been expertly tackled on

The Chimera 175 was introduced at the 1956 Milan Show, going on general sale shortly after.

1959 Chimera 175/250

250 in brackets where different

Engine:	Air-cooled ohv single, horizontal cylinder, alloy head, cast iron barrel, built-up crankshaft with roller bearing big-end, vertically split aluminium crankcases, unit construction.
Bore:	60mm (66mm)
Stroke:	61mm (72mm)
Displacement:	72.4 cc (246.2 cc)
Comp ratio:	7:1
Lubrication:	Gear type pump; wet sump
Ignition:	6 volt, battery/ coil; crankshaft-mounted 45w generator
Carburettor:	Dell'Orto
Primary drive:	Helical gears
Final drive:	Chain
Gearbox:	4-speed, foot change
Frame:	Spine type, with full enclosure
Front suspension:	Telescopic fork
Rear suspension:	Monoshock
Front brake:	Full width alloy drum
Rear brake:	Full width alloy drum
Tyres:	Front 2.50 x 17 Rear 2.75 x 17 (3.00 x 17)

General specifications

Wheel box:	54 inch (1,370mm)
Ground clearance:	6 inch (152mm)
Seat height:	30 inch (760mm)
Fuel capacity:	4.75 gallon (22ltr)
Dry weight:	269lb, 122kg (275.6lb, 125kg)
Maximum power:	9.25bhp at 6,500rpm (13.7bhp at 6,500rpm)
Top speed:	68mph, 110kph (74mph 120kph)

the Chimera. The cylinder is disposed almost horizontally with the light alloy head facing forward. Cooling air enters the open front of the die cast aluminium alloy engine cowl and impinges directly on the cylinder head. On test, no amount of hard riding produced the slightest symptom of overheating.'

Actually accessibility was good, certainly, considering that the engine unit, frame and swinging arm were totally encased. For example, the spark plug was positioned well forward in the combustion chamber and at a small angle to the cylinder, and thus could be removed with the spanner (albeit long) provided in the machine's tool kit, with the engine cowl still in place. When it was necessary to carry out more extensive work such as checking the valve clearances, the cowl could be removed by way of only two hexagon-socket screws, the cowl being located on four dowels. To the rear of the front single-piece cowl, were a pair of much larger engine panels

The cylinder head and barrel of the 172.4cc (60 x 61mm) overhead-valve horizontal single cylinder engine...

which encased the entire engine/gearbox unit from the base of the cylinder barrel backward. Again in cast aluminium, each panel was secured by a trio of socket screws. The carburettor tickler was accessible through a slot in the nearside panel which also incorporated a lockable tool box. The three way fuel tap protruded through a rubber sleeved hole in the offside panel. On the same side, without the need to remove the panel work there was instant access for both the contact breaker points and the engine oil dipstick, the latter being removed with a 14mm tubular spanner, again supplied in the machine's original tool kit.

The 'Motor Cycle' test found, 'The strangest of the early impressions formed by riders unaccustomed to small Italian sports models, concerned the riding position. By British standards, the handlebar setting is unusually low and flat and this involves a slight forward crouch apparently beloved by Italian youngsters. It was felt initially that the position, apart from inviting the criticism of 'pseudo racer', would place an undesirable strain on the rider's arms during a long run. Extensive

...and with the engine front cover in place.

La «Chimera» è rapidamente scomponibile in tutte le sue parti.

experience showed that any such stress was much less than anticipated; at the end of a day the rider experienced a mild ache in his elbows. No discomfort was felt on short runs.'

The riding position came in for considerable praise. The tester commented, 'The seat is well sprung, of ample length for two people and low enough for short riders to reach the ground easily. Relationship between seat and footrests furnishes an un-cramped knee angle for rider and passenger, yet footrest width is narrow enough to obviate grounding on corners. Passengers appreciate the upturned rear edge of the seat. All controls, both hand and foot operated, are conveniently sited.'

Special mention was also made of the early form of mono-shock rear suspension, 'Actuated by an upward extension of the fork bridge, a single non-adjustable shock absorber controls the movement of the rear fork. Spring and damping characteristics are very well chosen, as also are those of the telescopic front fork. At low speeds with a passenger, suspension might be considered a little firm, but at high speeds it cushions road shocks really well, without pitching or clashing and is remarkably effective, even when a passenger is carried. In that case the majority of bumps just seem to disappear uncannily beneath the wheels with no perceptible movement of the model.'

Even higher praise was lavished on the steering, the tester saying it was 'reminiscent of that of a lightweight-racing machine: light and positive straight ahead, yet almost seeming to possess a conscious desire for bends and corners.'

Engine starting was 'almost invariably achieved first kick' and it was found 'best to flood the carburettor slightly when the engine was cold and not at all when hot.' As for idling, this was 'extremely slow and regular and so unobtrusive that it was sometimes difficult to appreciate that the engine was running when the model was stationary in heavy traffic.'

But what of performance? The 'Motor Cycle' test sheet recorded a highest speed of 65mph, however a truer picture is obtained by reading the test report. 'On about half throttle the Chimera cruised contentedly for miles on end and indicated 90 kilometres an hour which, allowing for 10 per cent speedometer flattery is about 50mph. Slightly favourably conditions sent the speed up to 60mph without any increase in throttle opening. Given a clear run the little model would romp up most main road climbs without slowing to less than 45mph.'

Other features which aroused favourable comments concerned the 'smoothness and unobtrusiveness of the engine, light and positive gear change', plus the front brake which called for no adjustment after 800 miles of hard use, the snap fastening petrol cap and a detachable key for ignition, lights and horn. Criticisms were made of the rear brake lever, needing constant adjustment and overlong heel and toe gear change lever and on a model of such advanced design, an open driving chain was considered an unexpected anachronism.

The test concluded, 'Few will deny that large-scale advances in motorcycle styling are overdue. In addition to being one of the most enterprising conceptions ever, the Chimera is very cleverly planned in detail. It deserves every success and if it encourages the birth of other designs of equal merit it will have proved doubly worthwhile.'

However, even with all the publicity, and the fact that soon after a quarter-litre version (246.2cc - 66 x 72mm) was introduced, the Chimera proved a resounding sales failure, both at home and abroad.

Faced with a massive financial loss for his company, Bianchi was forced to retreat to his drawing office in an attempt to save the day. Expressly charged with making as few changes as possible in order to keep manufacturing costs down to the bare minimum, Bianchi produced a trump card, an undressing exercise, coupled with a redesign of the rear frame and swinging arm from single near-horizontal, single rear shock, to a conventional set-up with twin vertical units.

The results from these most basic of changes were to be absolutely incredible, with sales of the new naked 175 and 250 ohv horizontal singles rocketing.

The three colour schemes available for the Chimera during the 1957 model year were green, salmon pink and blue.

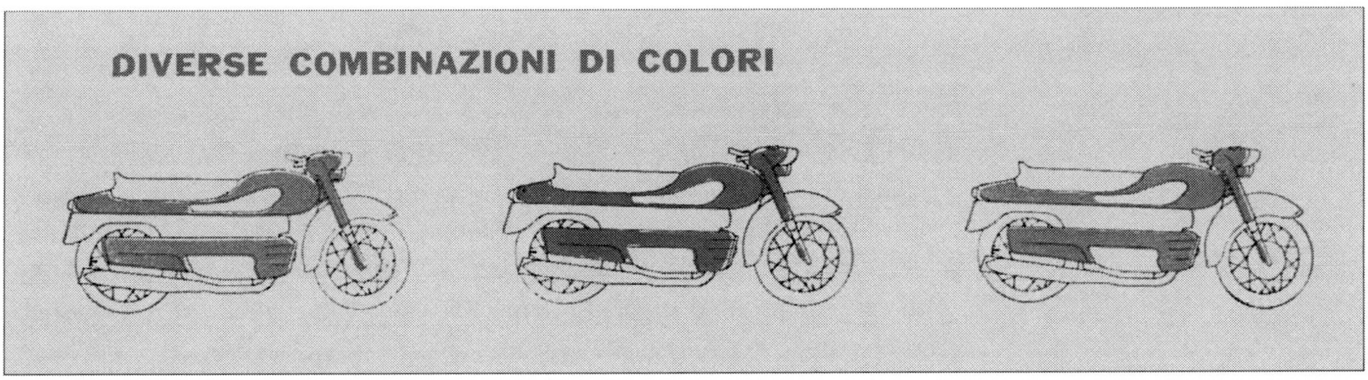

4 Ala Rossa-Ala Verde

By 1958 Aermacchi concluded that the futuristic Chimera would never sell in quantity. Ing. Bianchi 'undressed' it to reveal the Ala Rossa (175) and Ala Verde (250).

The Ala Verde (green wing) and Ala Rossa (red wing) were basically undressed Chimera models, with engine sizes of 242.6 cc (66 x 72mm) and 172.4 cc (66 x 66mm) respectively.

The two models had arisen as a result of appallingly low sales of the Chimera, which although widely publicised following its debut at the 1956 Milan Show had not proved as popular with the buying public as the press.

In 1958 and faced with a huge financial loss, Aermacchi's design chief Ing. Bianchi had been ordered by an unhappy management to resolve the problem. So with words of discontent ringing in his ears, Bianchi retreated to his drawing office to set about resolving it.

He did so in spectacular fashion. There were in fact very few major changes, even though the finished work looked entirely different. Bianchi, mindful of financial considerations kept retooling and manufacturing costs down to a bare minimum, while carrying out an undressing programme. This was coupled with a redesign of the rear frame and swinging arm from a near horizontal, single rear shock absorber, to a conventional set up with twin vertically placed assemblies.

So far, little more than a partial frame rework and new bodywork, Bianchi had produced what appeared to be, and were accepted as entirely new models. Not just that, but actually a whole series evolved. As well as the Ala Verde and Ala Rossa which were the sports roadsters, the Ala Bianca (white wing) 175 touring model, plus the Ala d'Oro (gold wing) 175 and 250 racing machines were developed.

Actually, the 175 series (Ala Rossa, Ala Bianca and Ala d'Oro) came a few months before the 250s, the former being launched in 1957, whilst the Ala Verde and Ala d'Oro 250s didn't arrive until the following year. This was because, in Italy at least, the market for 175 cc motorcycles during the late 1950s was much stronger than the 250 cc market.

Aermacchi's publicity department said the Ala Rossa 175 was 'for the sportsman'. They claimed 130kph (81mph), which was highly respectable for a pushrod single.

The 172.4cc Ala Rossa ohv horizontal single with alloy cylinder head, cast iron barrel and steeply inclined Dell'Orto UBF carburettor.

Dutch Aermacchi Club founder Jaap de Jong at Vandvoort with an Ala Verde 250 in 1961. Machine is essentially stock except for megaphone exhaust and small headlamp fairing.

Aluminium cylinder head and cast-iron barrel, Dell'Orto UBF carburettor and contact breaker points cover.

A journalist of the period described the bike, 'Layout of the Ala Verde is simple. Built in unit with the gearbox and inclined a few degrees above the horizontal, the engine is suspended from a tubular beam frame. Oil is carried in a finned sump at the base of the unit. Suspension is orthodox with telescopic front and pivoted rear forks. Short clip-on handlebars, knee recesses in the fuel tank and a narrow dual seat give distinctly racy lines.' This, of course, was in stark contrast to descriptions of the Chimera given out a few months earlier.

Whereas the Chimera had meant to appeal on its innovative styling above all else, the naked Aermacchi horizontal singles achieved success mainly because of other priorities such as sports and racing, the on-road performance, the powerful brakes and, above all, road holding ability contributed to the success story they became.

1962 Ala Verde, offside and nearside, a neat and purposeful bike.

1961 175 Ala Rossa/250 Ala Verde

250 in brackets where different

Engine:	air-cooled ohv single, horizontal cylinder, alloy head, cast iron barrel, built up crankshaft with roller bearing big-end, vertically split aluminium crank cases, unit construction.
Bore:	60mm (66mm)
Stroke:	61mm (72mm)
Displacement:	172.4 cc (246.2 cc)
Comp ratio:	8.2:1 (9:1)
Lubrication:	Gear type pump, wet sump
Ignition:	6 volt battery/ coil; crankshaft mounted generator
Carburettor:	Dell'Orto
Primary drive:	Helical gear
Final drive:	Chain
Gearbox:	4-speed, foot change
Frame:	Spine type
Front suspension:	Telescopic fork
Rear suspension:	Swinging arm, twin shocks
Front brake:	Full width, alloy drum, SLS
Rear brake:	Full width, alloy drum, SLS
Tyres:	Front 2.50 x 17 Rear 2.75 x 17 (3.00 x 17)

General Specifications

Wheel base:	52 inches. 1320mm
Ground clearance:	5 inches. 127mm
Seat height:	29 inches. 736mm
Fuel capacity:	4 gal. 18ltr
Dry weight:	242.5lb 110kg (247lb. 112kg)
Maximum power:	12bhp at 6700rpm (18bhp at 6700rpm)
Top speed:	76.5mph. 130kph (82mph. 140kph)

Another press extract from the time reports, 'Thanks largely to the unusually low centre of gravity given by the prone engine layout, road holding would do credit to a racing machine; braking power, too is more than plenty.'

Not only this but as the 'Motor Cycle' was able to report from a 1961 road test of the Ala Verde, 'As to running costs, the Aermacchi averaged almost 90 mpg over 1,100 miles of fast main road work and town pottering, a tank capacity of 3.75 gallons meant a range of around 300 miles'. A pretty impressive figure and unmatched by the majority of motorcycles of the time, regardless of engine size. Many small capacity Italian sports models might have had power aplenty at the top end, but this was often achieved at the expense of bottom end punch. Not

Another 1962 Ala Verde, but this time converted into a club racer with race tuned engine, 30mm Dell'Orto SS1 carburetter and remote float chamber, racing tank, seat, megaphone exhaust, alloy rims and much more.

so the Aermacchi which was helped by its long stroke 66 x 72mm bore and stroke. The 'Motor Cycle' 1961 test reported, 'The 250cc single completely escapes this compromise. Its beef at low revs would do credit to a 500 single while the same sort of surging power is on tap right through the speed range to a rear 80mph maximum.

'Like the Chimera, the Ala Verde/ Ala Rossa were equipped with 17 inch tyres, as against 18 inch for the Ala d'Oro racers. This of course aided the low nature of the motorcycles, with a seat of only 28.5 inches'.

Both the Ala Verde and Ala Rossa were very successful in production machine racing, winning events all over Europe. Several machines were also subsequently

A very nicely customised 1966 five-speed Ala Verde with some nice touches including Veglia racing tachometer, 2LS front brake, megaphone silencer, alloy rims and lightweight mudguards.

1964 factory photograph of the Ala Verde, very useful for restorers

converted into racers, for those who couldn't afford the pukka Ala d'Oro. In fact, in many ways, both the Ala Verde and Ala Rossa were more at home on the track than on the street. But in fairness, most of the bikes' weaknesses were shared with other Italian motorcycles of the era. These included poor lights (CEV), minimal rear chain guard, poor toolkit and thin chrome work, which deteriorated quickly. But against these faults the potential owner was offered class-leading performance, racer-like handling and road holding, powerful brakes and superb looks.

Berkshire enthusiast Maurice Stewart-Ashley, with his partly dismantled Ala Verde. This photograph shows the single backbone frame stucture; 17 inch wheels also contributed to a low centre of gravity.

1966 5-speed Ala Verde, with standard tank; front brake and saddle are non standard, as are rear light, silencer and footrests.

Jaap de Jong putting his Ala Verde to good use at Zandvoort, 1970.

This particular Ala Verde was originally owned by Syd Lawton's son Barry. First registered in 1964 it then passed to Brian Silver (now a classic Ducati specialist). Brian is seen with the bike during the late 1960s.

As with other Italian models, there was the traditional heel and toe gear lever on the offside for the 4-speed gearbox. The Ala Verde's ratio was, 1st 15.4:1, 2nd 9.3:1, 3rd 6.7:1 and top 5.5:1. Also there was a helical gear primary drive and a near-side placed kick-starter.

The Ala Verde/ Ala Rossa was usually a first or second kick-starter. The 'MotorCycle' described the recommended drill, 'Turn on the petrol (100 octane was always used) and ignition, depress the air plunger on the carburettor top and give a hefty swing on the kick-starter. Flooding the carburettor usually made starting uncertain. As soon as the engine fired, the air slide could be opened fully and a slow, reliable tick over obtained.'

A line-up of Dutch Aermacchi club members' machines awaiting their turn to go out at Assen, mid 1990s.

'Websterini'

William Marshall (Bill) Webster nicknamed 'Websterini' because of his close association with Italy, was the first British Aermacchi importer.

Born in Liverpool on the 10th July 1909, Bill became a very popular figure in racing and had close connections with Aermacchi, and also MV Agusta post Second World War.

Bill Webster began motorcycling in 1927 and rode for Liverpool Speedway until 1932. He was a member of the North Liverpool Motorcycle Club and it was here that Bill struck up a friendship with Les Graham, from the Wirral, then making a name for himself on OK Supremes at Donington and on grass circuits.

During the war Bill had worked at the Rolls Royce engine factory in Crewe, whilst Les became a salesman for the London based AMC concern. Both went racing at the first opportunity, and both made their mark in road racing. Bill Webster was one of the first to support 125 cc racing, still in its infancy in Britain.

On a sleeved-down 150 Royal Enfield, he won every race he started at Altcar (near Liverpool) in 1951-2. Bill had finished third in the 1947 250cc Ulster Grand Prix. That same year he had made his Isle of Man TT debut, finishing eleventh in the Lightweight event on a 250 cc Excelsior.

By 1953 Bill had purchased a new 125cc MV, Agusta (the first British private owner) and with it he finished sixth in the TT. He also came ninth on a 250 cc Velocette.

From then until his retirement from active racing in 1959 he became closely involved with Count Domenico Agusta, boss of the Italian firm. 'Motor Cycle News' once said, ' An extravagant brand of Anglo-Italian jargon, half spoken, half gesticulated, soon marked Bill as an ideal go-between for the Galarate factory in contracting British riders and officials.' Bill also had the satisfaction of being a member of a winning MV team in the 125 cc classes of the Isle of Man TT, Ulster Grand Prix and Dutch TT, as third man, first to Cecil Sandford and Bill Lomas and later to Carlo Ubbiali and Luigi Taveri.

He was involved in guiding Mike Hailwood during his early racing years and also sponsored several established or promising riders.

In 1960 he formed Italian Imports, based at Snow Hill, Nantwich, which acted as British distributor for both MV Agusta and Aermacchi. Bill entered several Aermacchi riders, including Alan Shepherd, Percy Tait, Stuart Graham and Brian Clark, all of whom were successful on the Varese factory's machinery.

Then on, Sunday 27th April 1963 tragedy struck at Mallory Park. Bill had returned from Italy the previous day, having visited MV and Aermacchi and the early season Imola road races.

At Mallory, there was an argument with officials regarding passes. Bill, who had suffered heart problems the previous year, collapsed and died, leaving his widow, Ethel, two daughters, and four sons.

The Aermacchi import side of his business was transferred to Syd Lawton and the MV imports ceased, but his motorcycle dealership in Crewe was to continue.

'MotorCycling', 1st May 1963, described Bill Webster as 'a craggy, tireless, humorous, entirely likeable character'.

5 Wet-Clutch Racers

Horace Crowder on Bill Webster's 1961 250, wet-clutch, 4-speed Ala d'Oro, Mere Hairpin, Olivers Mount, Scarborough.

As described in the previous chapter, a considerable amount of money had been spent on the design of the Chimera, which had debuted at the Milan Show towards the end of 1956. At first it seemed a waste of money until, Ing Bianchi, Aermacchi chief designer, undressed the machine and re-shaped it into a sports model. The result was the Ala Rossa (175) and Ala Verde (250).

The naked versions achieved such immediate popularity, that before long it was decided to produce a special version, which although not originally planned, would act as a running prototype of a full racer model for sale to customers at a realistic price.

The machine made its debut in the 1960 German GP, followed by another entry, this time for the Dutch TT at Assen. Here Alberto Pagani (son of the 1949 125 cc World Champion and later MV Agusta team manager, Nello) came home a highly creditable ninth. An even better result was achieved eight days later at the Belgian Grand Prix, where over the

1961, 250cc Isle of Man TT, Gilberto Milani, works Aermacchi; note leading axle front forks.

Bill Webster (seated on bike) first brought Aermacchi to Britain. He is seen here at Oulton Park, Spring 1962, with former 350cc World Champion, Bill Lomas, behind.

ultra-fast Spa-Francorchamps course in the Ardennes, Pagani and the pushrod Aermacchi confounded expectations by coming home a brilliant fifth in a star-studded field.

These excellent performances prompted Aermacchi management to consider building replicas of the Pagani machine for sale. Called the Ala d'Oro (Golden Wing), this was to prove a wise move indeed. The cost in Italy upon the launch in the summer of 1960 was 300,000 lire.

1960 was also very important in the evolution of the company for another reason which, in retrospect, was not entirely beneficial in the long term. In April that year, a fifty per cent stake was taken in the motorcycle division of Aermacchi by

Riding for the Pégase team in 1961 Piet Vervoegen was probably Belgium's finest 250cc rider of the era.

John Ashcroft winning at Scarborough in Spring 1962 on an almost 'straight-out-of-the-crate' 250cc Ala d'Oro.

One of Bill Webster's 250cc Aermacchis, Oulton Park, 1962.

the American Harley-Davidson marque. This came about for two reasons; the first being Aermacchi's own success in re-establishing its aviation business in partnership with the American Lockheed company, and the fact that by now, space was becoming a severe problem at the main factory in Varese which the aviation and the motorcycle divisions shared at this time.

The assembly plant for the three wheel delivery vehicles was also in the same town. On the outskirts of Varese, at the lakeside location of Schiranna, was the old seaplane factory with its slipways into the lake, and its lofty hangar, which provided storage space for both motorcycles and trucks. Finally, there were aircraft assembly shops and an airfield at Venegono and, a sideline, the Cantiere Navale di Apuania. Thisr was the shipbuilding division, with four slipways at Marina di Carrara, where ships from 2,500 to 16,000 tons were constructed.

The whole operation had become virtually unmanageable. For example, in the main plant motorcycle, delivery trucks and aircraft components were all being manufactured and worked on alongside each other! Obviously something had to change, and quickly.

Meanwhile, across the Atlantic, Harley-Davidson felt it needed a European plant with cheaper operating costs and the expertise to produce a range of lightweight motorcycles with which it could supplement its US built Vee

1961 Ala d'Oro

Engine:	air-cooled ohv single, horizontal cylinder, cast 3 ring piston, alloy head, cast iron barrel, built up crankshaft with roller bearing big-end, vertically split aluminium crankcases, unit construction
Bore:	66mm
Stroke:	72mm
Displacement:	246.2 cc
Comp ratio:	10.2:1
Lubrication:	Gear type pump; wet sump
Ignition:	6 volt battery/coil
Carburettor:	Dell'Orto SS1 30mm, with remote float chamber
Primary drive:	Straight cut gears
Final Drive:	Chain
Gearbox:	4-speed, foot change, close ratio
Frame:	Spine type
Front suspension:	30mm stanchion, telescopic fork
Rear suspension:	Swinging arm, twin shock absorbers
Front brake:	Oldani 2LS, full width magnesium hub
Rear brake:	Oldani SLS, full width magnesium hub
Tyres:	Front 2.75 x 18; rear 3.00 x 18

General Specifications

Wheel base:	53.5 inch (1,358mm)
Ground clearance:	5.5 inch (139mm)
Seat height:	29 inch (696mm)
Dry weight:	225lb (102kg)
Maximum power:	25bhp at 9,000rpm
Top speed:	110mph (177kph)

twins. So came about one of the strangest marriages in biking history, Aermacchi Harley-Davidson.

As part of this agreement, all two-wheeled production in Varese was to be transferred to the lakeside Schiranna location, leaving the main factory free to concentrate its efforts on aircraft production, which by now included the excellent MB326 advanced training and ground attack jet.

The sporting side of Aermacchi had not been forgotten in the agreement; and the result was, if anything, even more positive than before. On the sports production-racing scene, a 250 Ala Verde (Green Wing) won the important Monza six hour event, whilst in Britain another Ala Verde made its debut in the Thruxton 500 mile race where, ridden by the pairing of Arnold Jones and Horace Crowder, it led its class until forced to quit through transmission problems. However it made an impression, 'The Motor Cycle', referred to it as 'that fantastic Italian invader'.

A 1962 250cc Ala d'Dro, with original front mudguard which most owners saw fit to remove at the time.

These performances were just the beginning. At the final event of the year, at Oulton Park in October, 'Motor Cycle News' was able to report: 'In the 250 cc class there was no one to touch Percy Tait, and he romped away on Bill Webster's Aermacchi to win by a mile from Dan Shorey.' Tait was in fact riding Pagani's bike, arranged by Cheshire dealer and sponsor, Bill Webster (see separate box), who was about to become Britain's first Aermacchi importer. Known as 'Websterini' due to his close connections with the Italian motorcycle industry, he had raced various MV lightweights during the 1950s, and was a close friend of the Count, Domenico Agusta.

In Britain, Bill Webster had formed a separate company, Italian Imports of Snow Hill, Nantwich, Cheshire, which not only imported small numbers of Aermacchi roadsters (mainly the Ala Verde 250), but also had signed Alan Shepherd to partner Percy Tait for the forthcoming racing season. Back home, the Varese Factory supported not only Alberto Pagani, but also Gilberto Milani. April 1961 saw Alan Shepherd get off to a winning start at the Silverstone Hutchinson 100 Meeting; but his victory was achieved amidst controversy. Early leader, Mike

Hailwood, had been black-flagged for failing to notify officials of a machine change. The first batch of the Ala d'Oro production racers was built in both 172.4cc (60 x 66mm) and 242.2cc (61 x 72mm) engine sizes. But in Great Britain only the larger version was imported, and at over £450, it cost almost as much as a 500cc Manx Norton. Fifty Ala d'Oros were manufactured in 1961, of which less than ten were 175s. Webster's Italian Imports concern brought in about ten bikes, which included machines for his own use.

For 1962 the Varese works responded to the demand for the racer, thanks, in no small part to the performance of men such as Pagani, Milani, Shepherd and Tait. Aermacchi built one hundred and fifty Ala d'Oros for the 1962 season, thirty of which came to Great Britain.

Although expensive, the buyer got a competitive quarter litre machine, but these early over-the-counter racers were not as reliable as many would have hoped. Quite simply, under the thin veneer of a racer's dream was still a humble pushrod race-kitted roadster. And this is where the problems started. Although fast, some 115mph, these original long-stroke, wet-clutch racers were extremely fragile, proving costly for their owners. The motor was like a hand grenade with the pin waiting to be pulled out, and although there were some good race results, there were also many horror stories of disintegrating pistons, seized big-end bearings, snapped con-rods and dropped valves. In addition, the wet-clutch made starting difficult even with such a light machine. However, there were always one or two bikes to survive the race, and these were usually up at the front of the pack.

Silverstone International Hutchinson 100, April 1962. Aermacchi riders are Percy Tait (34) and Tommy Robb (44), Tait was 3rd behind Jim Redman (43) Honda, and Mike Hailwood (42) Benelli.

A 1962 250 Ala d'Oro with muguard removed and a dolphin fairing fitted, otherwise stock. Note the bracing tube between the swinging-arm pivot and the top suspension pick-up, compare with page 46. Syd Lawton came up with this modification in 1963 which was incorporated on the 1964 and 1965 production bikes from the factory.

Compared with the roadster version, the 1962 racer, coded DS, featured a downdraught head with 30mm Dell'Orto SSI carburettor and remote float chamber, alloy cylinder barrel, cast high-compression (10.2:1) 3-ring piston, a DS5 cam (a DS2 was used in 1961), several minor engine changes and close ratio gears (but still only four ratios). There was an alloy 18 litre fuel tank, single racing seat, a pair of racing Oldani brake hubs laced to 18 inch alloy rims (the roadster had 17 inch steel), and chrome plated shallow-taper megaphone. Other differences included exposed stanchions on the 30mm front forks and a glass

Stan Wright (138) leads Percy Tait - both Aermacchi mounted, at Scarborough, 1962.

Dave Habel's 1962 250cc wet-clutch machine at Snetterton, 1964.

fibre dolphin fairing. Cosmetically, the seat was a give-away to identify between 1961 and 1962 machines, but only on original photographs or brochures, as of course the seat could well have been changed.

Besides the pukka racing version, the Ala d'Oro was also marketed in roadster form for Italian Formula III events, but in this guise it was mainly sold as a 175. As a further point of interest, on the smaller machine a straight pipe was fitted in place of the megaphone on the 250.

John Hynes with his 1962/63 wet-clutch Ala d'Oro, Lydden Hill, Kent, 1967.

The 1962 production Ala d'Oro wet-clutch. The 246cc (66 x 72mm) pushrod engine delivered 28bhp at the rear wheel at 9,800rpm, with decent torque from around 6,700rpm.

Aermacchi-mounted Pete Slinger from Preston, one of the stars at the Wirral 100 Club's races,Oulton Park, Saturday 23rd March 1963. In his first Slinger was 3rd, behind Tom Phillips (Greeves) and Mick Manley (Ducati) and in his second ride finished runner-up to Manley.

At Grand Prix level, the 4-cylinder Hondas dominated 1962, but even so the Varese marque performed well. Particularly noteworthy was Pagani's fifth place in the Isle of Man Lightweight (250 cc) TT. Although there were in fact only seven finishers, this was still quite an achievement. Pagani's average speed was 85.5mph in a race which had seen both Shepherd and Milani also well placed before being sidelined, Shepherd on the fourth lap and Milani on the final (sixth) circuit of the famous 37.75 mile Mountain Course. Of the three riders, Shepherd was the fastest, lapping at around 90mph.

Other top six Aermacchi finishers in the classics that year were Pierre Vervroegen (sixth in Belgium); Stuart Graham (sixth in Ulster); Milani (fifth in Italy) and Dietrich (sixth in Argentina).

Besides the real production Al d'Oro racers, several Ala Verdes were converted for this purpose. These spurious examples can be easily spotted by their polished roadster alloy brakes, iron cylinder barrel, different front forks and other minor details. The Ala d'Oro was also sold in the United States as the Sprint CRTT (there were also CR and CRS dirt versions) and, confusingly, there were other roadster Sprints including the C and H.

By the end of 1962 it seemed pretty obvious to Ing Bianchi that reliability had to be improved on the long-stroke 250, so he carried out a thorough review of all the engine components which had proved troublesome. Following this, maximum output was increased to 28bhp at 9500rpm (previously peak output revs had been 9200rpm), and the adoption of a 5-speed gearbox enabled more effective use of available power. In addition, the reliability of items such as big-end, con rod and valves was greatly improved; the most notable changes being a forged piston, shorter connecting rod and an 'A' camshaft.

Behind the scenes, tragedy marred the new racing season when when after flying back from Italy, British importer Bill Webster collapsed and died at Mallory Park in late April. Webster, 53, had spent several weeks in hospital during 1962 with heart trouble.

And so came onto the scene a name known to all connected with the Aermacchi name, Syd Lawton. Lawton, a racer like Webster, was Managing Director of the leading Southampton dealership, Lawton and Wilson. For 1962 he had bought his racing son, Barry, an Ala d'Oro and, because of this, had become increasingly interested in the marque. In fact, he had recently visited Italy with Bill Webster, and later in 1963 became the new British distributor.

An article in the September 1963 issue of 'Motor Cyclist Illustrated' was written by the famous Italian journalist Carlo Perelli. 'Orders for these 250cc machines are flocking in from all over the world, two of their main attractions from the point of view of the lone head racer being the extreme simplicity of the overhead valve, single cylinder engine and the low cost (£290 in Italy). The Aermacchi, or Aermacchi Harley-Davidson as it had become, without any change in design or

Percy Tait at Brands Hatch in 1963.

The first two home in the 250cc race of the Southern 100 meeting in the Isle of Man, 11th July 1963, were Terry Grotefeld (51) and Jim Curry (37), mounted on near identical Ala d'Oro bikes.

A 1964 250cc Ala d'Oro, still with wet-clutch.

Some early Aermacchi racers featured Ceriani GP front forks. All had Oldani 2LS brake as standard.

From 1963 Aermacchi entered the 350cc class, at first these engines retained the wet-clutch as seen here in 1964.

appearance, since the American company linked commercially with the Italian, has a single beam frame from which the horizontally mounted engine is slung.' Carlo Perelli continued, 'This engine, with push rod operated valves, delivers 29bhp at the rear wheel at 9,800rpm and has excellent torque from 6,700rpm up, with quite good pulling power from as low down as 4000rpm. The inlet valve, which should have a clearance of 0.007 inches opens at 79° before TDC and closes 94° after BDC. The exhaust valve (clearance 0.009 inches) opens 98° before BDC, closes 66° after TDC. A 30mm (1.18 inch) choke Dell'Orto carburettor, with separate adjustable float chamber, is used and ignition is by battery and coil, with manual advance and retard from 50 ° to 55° before TDC, a Lodge RL50 sparking plug being fitted as standard. Wet sump lubrication with gear pump is employed. When first introduced (1961 and 1962) the Aermacchi had a 4-speed gearbox, but now it has 5-speeds as normal equipment, the internal ratios being 2.263, 1.757, 1.378, 1.126, 1.00 to 1. A 16-tooth gearbox sprocket is standard, with an 18-tooth if the machine is equipped with a dolphin fairing, in both cases the rear wheel sprocket having 48 teeth. Fuel tank capacity is 13 litres (2.9 gallons). The 18 inch wheels are fitted with 2.75 inch front and 3.00 inch rear tyres, front brake is twin leading shoe, rear is single leading shoe. Total weight of the Aermacchi is 207lb, the maker's standard fairing adding a further 9lb.'

Colin Fenton, brother-in-law to Andrew French, 4th man home in the 1965 Manx Grand Prix on his wet-clutch 250, his average speed was 86.82mph.

Terry Grotefeld, Manx Grand Prix 1964.

Other details of the 1963 Ala d'Oro 250 racer include gear primary drive, with secondary gear for driving the camshaft contact breaker and oil pump; twin coil springs for the valves, the inlet valve being of 37mm diameter, the exhaust 29mm, the dome shaped piston (new forged) featuring two compression rings and one scraper. The small end was a plain bush, the big-end caged rollers, the light alloy cylinder barrel incorporated a tunnel for the pushrods, whilst there was an austenitic iron liner. The frame design was essentially the same as the Ala Rossa sports models, but the rear shock absorber damper units were in electron for lightness and had no fewer than five settings for load variations.

1962 saw an interesting racing prototype appear. This was an over-size 250, and had first come about to allow the company to compete in 500cc motocross events at the end of 1961. It employed a one-off piston and cylinder, grafted to a standard long stroke Ala d'Oro bottom end. The square 72 x 72mm bore and stroke figures gave 293cc.

At the end of 1963 it was announced that not only was the 250 Ala d'Oro to be updated once again, but it would be joined by a full 350 version. The smaller bike was to become a short-stroke with new measurements of 72 x 61mm. It was to have needle roller cam bearings and a choice of 'A', 'AF' or 'F' (350) cams.

As for the new 350, this had a capacity of 344cc, with bore and stroke measurements of 74 x 80mm. Using the 250 as a basis, but with the stroke

Pilot's eye view of the 1964 Aermacchi.

Spalding, Lincolnshire rider Brian Johnson with his 1963/64 wet-clutch 250, CRMC Cadwell Park, April 1983.

increased by 9mm, whilst bore size was fixed at 74mm- 2mm up on that of the latest 250, the 350 was only used by factory supported riders in 1964. One of these '350' motors had been supplied to Lawton and Wilson back in July 1963. This has 39mm inlet and 29mm exhaust valves, and 9.5:1 compression ratio. Syd Lawton himself rode it at Thruxton for the August Bank Holiday meeting, mounted in a standard 250 Ala d'Oro chassis. He found the top end performance disappointing, but the acceleration was good. Development work continued on the prototype engines both in Italy and Britain. A number of these were 4-speeds, but most were built with a 5-speed transmission. Other details included wet-clutch: 30, 32 or 35mm carburettors, forged piston, long con-rod, 'A', 'AF' or 'F' cam and 28.5bhp at 8,000rpm. The compression ratio finally settled at 11.6:1 after several stage increases.

The highlight of the works development programme with the 350 was a pair of fourth places, by Gilberto Milani at the West German Grand Prix at Solitude and by new signing, Renzo Pasolini at the Italian Grand Prix. Yet another fourth place, at the Sachsenring in East Germany, was the best a 250 could do in the classics that year, piloted by Milani. In September 1964 a 250 Ala d'Oro again took overall victory in the Monza six-hour race, with Alberto Pagani and Guiseppe Visenzi as co-riders.

The American market was becoming ever more important, with 75% of the Varese factory's output now crossing the Atlantic. The most popular models were

Mike Attrill with his Lawton-converted 250 Ala Verde roadster, Brands Hatch, 12th October 1985. Front brake is a Fahron unit. Rider and bike later won the Kennings Classic Group, Period Championship.

the various Sprints, one of the American favourites being the H, a street scrambler with abbreviated mud-guarding, smaller tank and a one-off saddle. The Sprint H was renamed the 'Scrambler' in 1964, with a bigger carburettor, higher compression ratio, revised exhaust and more power (25bhp).

America also saw a series of record attempts using a specially prepared short-stroke 250 Ala d'Oro road-racing engine. After being thwarted by rain earlier in

Dave Lecoq's 250 Aermacchi at a CRMC meeting, Cadwell Park, August 1985.

The 1963 pre-production 350 Ala d'Oro originally owned and raced by Syd Lawton is now the property of Oxford enthusiast Andrew 'Froggy' French.

1964, Harley-Davidson Works rider, Roger Reimann, managed to secure a speed record at the Bonneville Salt Flats in early October. Enclosed in a 14 foot long shell, the record breaker averaged 156.24mph (on petrol) for the flying mile, and 156.54 for the kilometre. Subsequently, the records were approved by the AMA (American Motorcycle Association) but not the FIM (Federation Internationale Motocycliste), since no recognized observer was present. Hence the speeds constituted American records only.

In Britain, during 1963 and 1964, the 250 Ala d'Oro had a new challenge on the track. This was a new breed of single-cylinder, two-stroke racing machine from Greeves, Cotton and DMW. This story had really begun at the Earls Court Show in late 1962, when various prototypes from these three manufacturers were launched. Not only this, but the Wolverhampton-based engine specialists, Villiers also debuted a brand new engine, the Starmaker. This was used in both the Cotton and DMW designs, whereas the Greeves used a Villiers 37A motor with special Alpha crankshaft. All three presented a serious threat to the Aermacchi in Britain, both on speed and more so price. There were also new bikes from foreign manufacturers, in the shape of the Bultaco TSS and the Yamaha TDI twin. But these were very much in early development, and suffered both on reliability and cost considerations.

In the early days, the Ala d'Oro had suffered more than its fair share of mechanical weaknesses, however by 1964 this state of affairs was improving.

The factory's European racing support programme meant that a team of official riders acted as works riders-cum-testers, performing a similar role to that

Nicely restored 1964 wet-clutch Ala d'Oro. Although parts are more expensive than the road-based Ducati singles, the Aermacchi has proved popular in classic racing since the CRMC was formed in 1979.

Wet-clutch 350cc engine with Dell'Orto SS1 carburettor and Amal Matchbox float chamber.

undertaken by Norton team members from 1955, in developing the Manx dohc singles. But at this time, unlike Gilberto Milano and new boy Renzo Pasolini, Alberto Pagani rode a semi-works machine with a one-off tubular space frame that he had constructed himself. The Aermacchi factory policy called for these works supported riders to concentrate on the important international meetings in Italy and abroad, as well as the classics, to enable Ing. Bianchi to further improve the over-the-counter production racers for the following year.

Particular works development (as opposed to production versions offered to the public), in late 1964 and 1965 centred around improvement to the strength of the gearbox, a dry instead of wet-clutch, and measures aimed at improving both the general performance and reliability of the Ala d'Oro two-fifty and three-fifty factory entered machines. There is no doubt that this race tested approach was to pay off later, notably from the 1966 model year of the production 'over-the-counter' bikes, the story of which is covered in Chapter 8.

The Clark Way

Tommy Robb (12) and Brian Clark, both Aermacchi mounted, Mallory Park1962.

My first sight of an Aermacchi was at the Scarborough International in 1960. Bill Webster's first 'Macchi, came on loan from the factory, for Alan Shepherd. Hailwood on his Mondial was the obvious favourite. However Mike's carb came loose, Alan won. I finished 6th on my 175 Ducati Formula III, and Mike was 7th, the only time I ever finished ahead of him! After Scarborough I remember Percy Tait riding and winning at Aintree and Oulton on the same bike. I don't remember many Macchis during '61, though I shared starting positions with Alberto Pagani on a Macchi at the TT that year. By '62 there were a few more and I realised that my Ducati, though now bored out to 198cc and quick, was going to be regularly beaten. Since buying it from Bill Webster in 1959 we had become friendly, but I was still surprised when he approached my sponsor, and future father-in-law, Tom Lambert asking to let me ride a 'production' bike with Stuart Graham at the 1,000 km race at Silverstone on 19th May that year. It was decided I would ride an Aermacchi at Mallory three weeks before the Enduro, 29th April, to gain some experience on one. With 2nd in the heat and 3rd in final I was well pleased. Two weeks later was my last race on the Ducati with a 2nd at Brands.

At Silverstone the bike was not very good in practice, it was really a racer with a silencer and it didn't like it one bit. Stuart started the race in poor weather, and the bike broke down in that first session. So I waited all day for nothing. In those days you couldn't get out of the paddock until racing was over, certainly not while an Enduro was going on. The other bike was ridden by Alan Shepherd and Tommy Robb. It did not last long either. The day after (19th May) I had a pukka Ala d'Oro racer for Snetterton. I was second. At the post TT Mallory on 10th June I was 2nd in heat and 5th in the final (250cc) and Bill let me ride a one-off 285cc model in the 350 race. Being small, it was better for me than the Manx I normally used. But it was a sod to get started. It had been fitted with an advance and retard lever to get it started. Its valve timing was obviously not right because after three laps the valves got themselves entangled with the piston. I think this was the last (and probably first) time it was run. It would only rev to 8,000. The following day, a Bank Holiday (11th) was Aintree. I dropped the Norton in practice, due to a flat tyre. I was walking away with the 250 race, but as I started the last lap the bike died. The condenser wire had pulled out and not for the last time.

At Scarborough on 15th June, was the 'Cock o' the North' event. Leading heat one into the last lap and guess what. The condenser wire pulled out. This was the bike that Alan Shepherd had ridden in the TT with no problems. It still had its 'works only' Dunlop triangulars on. And what tyres they were for a circuit like Oliver's Mount. Horace 'Crasher' Crowder won the televised final the following day on another of Bill's bikes.

Two days later it was Prees Heath in Shropshire, where I won both heat and final, beating team mate Stuart Graham in the final with another Macchi ridden by Eric Cheers, in 3rd. Eric's was a converted roadster I think.

July 8th at Brands, the engine seized, lying 3rd with two laps to go. Some consolation was a 6th in the 125s on Bill's Bultaco.

July 29th, Snetterton, saw awful weather, and I dumped the 250 at Sears, though managed a 4th on a Bultaco in the 125cc.

August 6th saw the annual British Championship event, at Oulton Park. No chance of catching Minter and Redman on Honda fours. With a not too good bike, I was happy to finish 8th.

I ought to mention that I didn't actually have a bike. I would drive from Retford to Crewe on a Thursday or Friday night, collect a bike, use it for the weekend and return it Monday. Sometimes, if Bill was attending a meeting, he and his right hand man, 'Big Jack' Hedge, would bring a bike for me.

The following week was the Ulster GP. Stuart and I were entered and Bill and Jack would be attending. Unfortunately Bill was very ill and couldn't make it. I was also very upset because Bob McIntyre, with whom I had become friendly, had crashed at Oulton and died later from his injuries. I must say that this incident shook me.

With bad weather at Drundrod, I was not happy, I hadn't really managed to learn the circuit. However, I was lying 6th after seven laps. I was ahead of Stuart when once again, the condenser wire came out. Anyway Stuart went on to pick up his first World Championship point.

My next meeting was my 'local' track, Cadwell, where I forgot to switch the ignition on! I was lucky to get back up to 8th.

September 21st -22nd brought me to my favourite circuit, Scarborough. I got a 2nd in my heat and a good scrap in the final with Dan Shorey for the second year running. I knew that I could pass him anytime so I was gutted when the big-end broke up at the Esses on the last lap. Achieving the fastest lap was no consolation I can tell you. 7th October I was 4th at Brands.

During the winter I visited Bill to discuss the following season. He had done a deal with the factory tester, Gilberto Milani, to have three bikes with factory specification engines. In return he would provide Nortons for the TT for Gilberto. These three bikes would be for Stuart, Percy Tait and myself.

The 'specials' didn't arrive in time for the first race, at Mallory 31st March. I was 4th in heat and 6th in the final on a production bike.

The following week, 6th April, was the Hutchinson 100 at Silverstone. One special had arrived by air and this was given to Stuart. Redman (Honda 4) and Hailwood (Ducati Twin) were 1st and 2nd, I slipstreamed Ralph Bryans (Purslow Benelli) in third place across the line to get myself 4th while Stuart, 100 yards behind me, was lapped as he approached the line.

Good Friday, 12th April, I was at Brands Hatch on my 'new' bike., a bike that would stay with me, which I could tailor to fit. Derek Minter on the Cotton flew off the line as 2-strokes did, but by the end of the lap I was up there with him. I settled about 7 - 8 lengths behind because these faster 2-strokes had a habit of seizing, fast, and I didn't want to ram him from behind. Sure enough, on lap three as we left 'Surtees', a little black line appeared from his back wheel and I was past and cruising to an easy win. It was a long drive home to Redford with car and trailer, no motorways in those days and hardly any dual carriageway either. After work on Saturday the boss loaned me a second-hand van for the rest of the weekend. On Saturday

night it was a four hour drive to Snetterton.
Easter Sunday, in foul conditions I got a 2nd at Snetterton behind Ginger Molloy on a Bultaco. Through the night we travelled to Oulton. Actually we stopped off in Redford and swapped to the car, the van was useless. It was an International permit event and another 1st.
As they walked towards the dummy grid together, Big Jack asked my Dad what were we playing at? We had been changing sprockets and panic had set in with the other five riders for whom Bill has supplied bikes. Dad didn't tell him that we always left it until the last minute to see which way the wind was blowing before deciding what gearing to use. As the wind would be following down towards Knickerbrook we upped the gearing. I always remember reading a John Surtees article years earlier in which he stated that with five gears always gear top for the fastest bit and you would be OK elsewhere.
Having been elsewhere for practice on the Saturday, I was on the back row for the race. Bill gave me a good tip. He told me that from the Starter picking up the flag and mounting his little rostrum, he would drop the flag exactly fifteen seconds later. As the rider who should have been on the inside of the back row was a non-starter I moved to his position. At 14 seconds I was moving and by 16 seconds I was away and through Hall Corner in 3rd place. I needed to try something cute in a field of almost forty!
At the end of lap one, Jack turned and said, 'That will probably be the only one running at the end.' Not quite true, but there were only two, Derek Woodman and myself. The others had lowered their gearing and were revving the socks off their motors. Stuart's exhaust fell off, possibly he hadn't tightened the bolt enough. Ralph Bryans on Fron Purslow's Benelli blew up and I was dodging bits of valve and piston down to Knickerbrook. There then followed a game of cat and mouse with Robin Good on the Ariel Arrow. He led for most of the race, an 18 lapper, but on lap 17 I squirted my bike and that was that, 1st place.
Bill was delighted. And prize and bonus money was something I had only dreamed about. However the next meeting two weeks later at Mallory, Sunday 27th April, was to be one of the saddest days of my racing life. Before I had even arrived, Bill Webster, after an argument with officials over tickets, had collapsed and died. We practised, but weren't sure whether to race or not. Big Jack returned from hospital and told us that Bill's wife said we were to race in his memory. I managed 4th in heat and final, the leading Aermacchi.
Syd Lawton, the Southern distributor for the bikes, took over the franchise. Initially he said he couldn't support me as well as his son Barry. I persuaded him to let me take part in two events I was already entered for. At Brands, 12th May, I finished 2nd and at Snetterton, 26th May, I was 1st. As it was to be my last race on the bike, I had given the bike its head and I finished miles in front of the second placed man. After Snetterton, the late George Collis, who also lived in Southampton took the bike back to Syd. I enclosed an envelope with his percentage of what I had won. On Tuesday he rang me at work and said I could have the bike back for the rest of the year. Barry was using the Stuart bike as Stuart had rides on 350cc Ajay and 500cc Matchless machines on which he started to make a name for himself. I don't know where the third one went.
The following weekend, 3rd June, was an International at Brands and I won again. Mick Manley on a production Ala d'Oro must have thought he was going to win. But I knew differently, Once again a 2-stroke shot off the line first, Griff Jenkins on Charles Mortimer Snr's Greeves. At the end of lap one I was up there. Griff was all over the road so I kept just far enough back. Lap three he dumped it at Paddock, though the way it was written in the book 'We Went Racing' they

obviously had a different view. I looked back as I rounded Druids and saw Mick. I eased back and let him take me into a corner. As he did so, he knew that I had let him. For a few laps he led but coming up to lap a back-marker into Dingle Dell, who took a weird line, I suddenly found myself back in front. Four laps later Mick passed again. As he told me afterwards, 'Damn, the bastard's conned me again!' And so we stayed this way until the last lap. Up through Hawthorns I opened up the throttle and streaked by and was a comfortable distance ahead at the finish. That wasn't the end though, because the photographers played hell with me because I had spoilt a good picture by not overtaking through Clearways! As I said at the time, 'I couldn't have guaranteed winning from there.'

Two weeks later at the post TT at Mallory. I was 3rd in heat and 4th in final.

This was followed by the June Scarborough meeting where a broken con-rod in practice, as I was braking for Mount Hairpin, put paid to any chance of a race, let alone a win.

With the holes welded up, I was off to Brands, 14th July. Tommy Robb won and I was 2nd.

At Mallory I came 5th, and at Snetterton on the 28th I was 1st. It was an eight-lap race and I was 4th at the end of the first lap and cruising. I decided to dive up the inside into the Esses (the old circuit this) only to hit oil dropped by a sidecar outfit in the previous race. I flew up the slip road, turned round and was not allowed to re- enter the circuit until everyone else had gone by.

Rejoining last of forty, I got the bit between my teeth. Entering the last lap I was 5th. Onto the Norwich Straight, I swept into the lead. The flag-man missed me, so I did another lap just to be sure. Trevor Barnes (Guzzi) said afterwards he had seen me coming as we entered the Norwich but I didn't give him a chance to get into my slipstream.

Then it was the British Championship at Oulton. Tommy Robb, Luigi Taveri and Jim Redman on Honda twins shot away. I stayed reasonably close to them for half the race and then got a misfire that lost me a thousand revs. I finished 4th, about 30 seconds back, and lapped 5th man Ginger Molloy (Bultaco) twice.

Guy Tremlett, Competition Secretary of Bemsee (BMCRC) reported in the club magazine on the exploits of Club Members. 'Clark was obviously trying very hard to keep up with the Hondas because through the fast Knickerbrook bend, sparks were flying from underneath.'

Actually my Dad had noticed that the exhaust was well scraped but he also pointed out that a lug under the frame was also without paint. I believe that the Ala d'Ora and the Ala Verde frames were the same and this lug was for the centre stand on the roadster.

1st September was Brands and in appalling weather I was brought off at Druids on lap 1. Not a lot to say.

8th September at Snetterton an exhaust flange gasket blew and the performance was diabolical, eventually finishing 13th with the exhaust hanging off at the front. Somehow I was not black-flagged.

15th September at Cadwell, I was in pole position with Germany's Gunter Beer on a factory Honda 4. I messed the start and was last away. I was up to 6th on lap two but the con-rod broke as I was overtaking Mick Chatterton (Aermacchi) through the Gooseneck. That was hairy. Cadwell, my local circuit was not good to me, and my favourite, Scarborough, was not kind either.

With the holes re-welded in the crank-cases, I went to Scarborough the following weekend and the track had just been resurfaced. My goggles were shattered by a flying stone during the lap I was timed on for grid position. I was on the back of the grid for the heat and I cruised through to 3rd. With the fastest from the two heats I expected to be 4th on the grid, but in fact I was last, 18th. I was told that practice times were used in the finals (yes, I realised afterwards that this was the rule, but a bloody stupid one! What is the point of working out who were the fastest 18 and then reverting to the practice times? Race secretary Claxton threatened to exclude me if I argued any more, particularly when I asked If he'd have done the same if it was Geoff Duke, John Surtees or Mike Hailwood.

I tried to anticipate the start but the guy in front stalled so I hit him and fell off, right in front of TV cameras. I sorted myself out and set off in a rage. By lap five I was 3rd and catching Beer on the Honda 4, but couldn't see Tommy Robb who was on a twin. I was pushing Beer hard when at the Memorial he ran out of road. I was through, after Tommy. I was gaining, gaining, but the misfire returned and a thousand revs loss at Scarborough! I'll leave you to guess my thoughts. But amazingly I was still gaining. Into the twelfth and last lap I was 100 yards behind Tommy. I got alongside him at Mountside but had to give way or I might have had us both off. At the line we were 25 yards apart on the final circuit. No disrespect to my friend Tommy, but I had lapped 75 yards quicker than him. Claxton and I had more words. I hadn't been helped by a fuel tank that split around the filler neck. Still I had my 15 minutes of glory. Apparently the cameras hardly left me, though to this day I have never seen a replay of it. (Does anyone know what happened to ABC TV's sporting archives?)

Speaking to Tommy afterwards he said that Geoff Duke was signalling to him that he was miles ahead. He realised that it was me behind and gaining and so he had pushed the Honda to the limits of its handling and braking, which did not match the 'Macchi.

A member of the Lincoln Club, Terry Organ, was marshalling half along the back straight. As I passed him for the first time with the rest of the field already out of sight, the expression on his face told me what he thought of me. But as the race progressed and I moved through the field, he became more frantic and urged me towards the front. I am sure he helped. I wouldn't have been able to face him anymore if I had not recovered.

An overnight journey to Brands, another tank fitted, another engine fitted because we couldn't work out the misfire. This engine was worse that mine even without a misfire and I was lucky to finish 9th.

29th September, the Race of the Year meeting at Mallory I was 6th.

The following weekend was my wedding day so I missed a couple of meetings. The weekend at the end of our honeymoon was to be my last race, Brands Hatch. Meanwhile, my brother Geoff and my dad had taken the bike to a local airfield to find this elusive miss-fire. My brother was a good mechanic. He had suspected it was fuel, and was determined to identify the problem. The 'special' engines developed as much power at the back wheel as a production bike did at the crank. My bike revved to 10.800 against 9.800 for a standard engine. The problem reared its head during longer races because the spring of the contact breaker weakened as it got hotter. It was so simple! If only we had changed the points for every meeting!

As I said earlier, Bill allocated the 1963 machine to me, and Syd allowed me to continue using it. The seat was lowered, I was one of the few who retained the heel/toe gear lever, which was tailored to my size 6 boot, allowing very late downward changes without upsetting the bike's balance. I have already mentioned that the bike developed as much power at the back wheel as a 'production' bike did at the crank. The head and cam were different. We had fitted Girling rear shocks and we changed fork oil (somewhere I still have the official engine certificate signed by Milani).

Whilst with Bill I had to use Dunlops but was never totally happy with them, they just didn't seem as good as Avons, which felt good immediately. From mid-year Ferodo allowed me to have 'Works' green linings on the front Oldani. After initial snatching problems, we gave it a very long leading chamfer. What a brake that was. An extra 50 yards flat out before braking into Snetterton's old hairpin. In practice and at the meeting I slipstreamed Ian Goddard on a quick 7R. He sat up and braked and I flew past, still flat on the tank. He was convinced I would be on the deck around the corner, but I was disappearing over the rise. I'm very grateful to Ron Herring, who worked for Syd and re-built the motor every time it blew up.

At the Race of the Year, my Dad had walked the paddock getting various friends to say a few words onto tape recorder, ready for the following week's wedding; Tommy Robb, Syd Lawton, Dan Shorey to name a few. Dan implored me not to give up. I dithered but in December I discovered my eight-year-old leathers didn't quite fit anymore, so I quit. In February I received notification from Brands, who owned Snetterton, Oulton and Mallory that they would double my appearance money. Syd also rang and said he had a 250 and a 350 waiting. Too late, I was committed. If Syd had just hinted that he had tested/ raced the 350 the previous year at Thruxton, whilst I was finishing 4th behind the Hondas at Oulton, I would maybe have carried on.

Who knows, perhaps I would have done well, but like a good many more in that era, I may well have finished up prematurely in a box. I had three racing heroes, Les Graham; killed at the TT; Bob Mac killed at Oulton and Mike Hailwood in a car crash.

I have some great racing memories, particularly with the Aermacchi in 1963. And through Mick's books I can relive those days.

Brian Clark

Bill Webster (centre) with Brian Clark at Oulton Park on 15th April 1963 - sadly Bill was to suffer a fatal heart attack only two weeks later at Mallory Park on Sunday 28th April.

Moto Grotto

Terry Grotefeld took delivery of a new 5-speed machine on 7th October 1963.

I began racing in March 1959 when my first racing entry was accepted. I rode a 175cc Ducati Silverstone at the Brands Hatch Good Friday National Meeting. I was on the second row behind Mike Hailwood, Dave Chadwick, Dan Shorey and Jim Baughan (125cc British Champion of the year before). It was very cold and wet but I had a good start and was away in 6th place. The track was slippery and I only had CEAT road tyres so the bike was sliding around a lot, but being young and naïve I did not seem to notice, and actually caught and passed Jim Baughan. This did not last, he easily passed me again and pulled away. The race was a win for Dave Chadwick with Mike just behind him. I finished 6th and earned £6 in prize money. My father was so pleased he doubled it and I ended up with £18. Two days later I went to Crystal Palace thinking I could do it all again. I crashed heavily into the Glade and wrote off the front forks, which cost £18 to replace! I learned rapidly racing was not as easy as I thought, and not cheap. The little Ducati gave me many good rides and several wins over two years. Then I realised it was time to move on.

It was during the 1960 season that the first Aermacchi production racer was seen in this country. The legendary Bill Webster had imported this superb racer, and it soon became very clear to everyone, including me, that this was the way to go. Near the end of the 1961 season I asked Bill if he could get me one and was delighted to hear he was expecting a few from Italy, and would reserve one for me. This bike arrived in October and cost £300. I believe it was only the 5th one to come over. I eagerly drove to Nantwich to collect the Aermacchi from Bill's workshop where it was being finished off. The fairing had not come with the bike, so I used a Manx Norton number plate. I took the Aermacchi to Brands for a running-in practice session. The Aermacchi was much faster than the Ducati, but this was short-lived. A broken rocker arm put an end to the session. A new rocker was sent down from Bill and the bike was prepared for its first race at Snetterton, the end of season National. I had a very good start and some great dices and finished 5th, much to my delight, and ended up 'in the money'.

The first Aermacchis had a four-speed gearbox with wet-clutch. The long stroke engine had a bore and stroke of 66 x 72mm. Weaknesses in engine design soon became apparent. The racer was based on the roadster and the parts that failed were mainly things like big-end bearings, connecting rods, rocker-arms and spindles and worst of all dropped valves which would wreck the head, piston and barrel. My bike suffered many of these, including a catastrophic blow-up whilst in the lead at Crystal Palace, on the last corner of the final lap. An inlet valve head dropped destroying the head and barrel, I whipped in the clutch and coasted over the line finishing around 7th but with a very dead engine. Because I worked as a heat treatment engineer, I was soon able to find reasons for these failures. The Italian hardening and tempering process left something to be desired. Some of the components were too hard and brittle, I was lucky to be in a position where I was able to re-heat treat these and make them more reliable.

I rode this Aermacchi in the Southern 100 in the Isle of Man the following July (1962) finishing 3rd. I had many great rides on the four-speeder including a win at the August BMRC meeting at Silverstone and another victory at the Snetterton October races, a very pleasing way to finish the 1962 season. Later on we heard the 1963 engines were to have five speeds so I decided to sell the 1962 bike at the end of the season to Peter Heaney.

Very sadly, Bill Webster died after the 1963 season started, and the Aermacchi agency was taken over by Lawton and Wilson of Southampton. I ordered a new five-speeder from Syd Lawton, then southern Aermacchi distributor, and took delivery of this bike on 10th April for a cost of £433. 2s. 6d. This doesn't sound much now but was a very large amount then in comparison with our wages at the time. Interestingly, I managed to persuade Peter Heaney to have the new 1963 small seat, so I could keep the 1961 larger component, which was used by me over future years.

The new 1963 5-speed bike was much better for short twisty circuits such as Scarborough, Lydden and Crystal Palace. The closer ratio gears made it possible to keep the engine in the best rev, range. I do not think that the long-stroke five-speed bike was much faster than the four-speed model in a straight line.

I was due to take the Aermacchi to the IOM for the Southern 100 in July but it had dropped a valve and damaged the head at a previous meeting. I had arranged to meet Syd Lawton at his home near Southampton to get the essential spares. Syd had just returned from the Aermacchi factory in Varese and was resting till early evening. After a splendid meal at the Lawton home, Syd and I departed for my North London home at Mill Hill. I got home around 10pm.

As I was due to catch the midday ferry to the Island the next day, I decided that I would just slip the new piston and barrel on the crankcase and then fit the repaired head. Shock horror, the top diameter of the cylinder was bigger than the recess in the head. The barrel would have to be removed and the spigot machined about 3mm. By 11pm I had set the barrel up on my lathe. The liner was very hard and difficult to machine. Suddenly the lathe tool dug in and the brand new cylinder flew out of the chuck and up into the air! With a wild desperate leap I caught it. I then decided to make a proper support for the barrel and so it was machined correctly until the head fitted perfectly. It was now much too late to finish the engine so I called it a day and went to bed. Next morning I loaded the van and set off for Liverpool. Once the car was on the ferry and we were at sea I sneaked down to the van and proceeded to

Flat out down Bray Hill during the 250cc Manx Grand Prix, September 1964. Terry finished 3rd, averaging 83.08mph for the 4 lap, 150.93 mile race.

rebuild the engine. We arrived at Douglas and set off for Castletown but had missed the day's practice sessions so I had to wait until the next day, which was race day, but there was one more practice period in the morning and I managed to qualify. A few hours before the racing was due to start I paid a quick visit to the Fairy Bridge to ask the 'Little People' for a bit of luck after all the problems of the previous weeks.

Just before the 250 race, all the riders set off from the paddock on the lap of pageantry and somehow I arrived at the start-line and then the engine stalled. A quick look at the carburettor showed that the small wire retaining the pin that stops the castellated ring undoing had vanished, allowing the throttle slide to come right out. Just as the starter was raising his flag I yelled madly for him to wait and then turned to the nearby grandstand and asked if any lady in the crowd had a long hairpin I could have to fix the problem. One was quickly found and passed to me, I fitted the pin, bent the end over and got ready to start. As the flag dropped I made a very quick start and was second away right behind the favourite for the race Alan Dugdale on the ex-Mike Hailwood ex-works 250 Benelli single. As we rounded the first corner, Alan was pulling away quite quickly until suddenly there was a puff of smoke from his megaphone as an inlet valve dropped in and he instantly slowed and pulled over, I shot past and into the lead where I stayed until the finish, getting the fastest lap of the race... Thank you little people. That race was definitely the best one for the 1963 Aermacchi and me.

The early Ala d'Oro racers (1962 models) had a

Chin on tank at the International Race of the year meeting, Mallory Park, September 1964, finished 9th.

fragile big-end assembly in as much as the connecting rod only had one rib running round the big-end eye, as I told you earlier they tended to be hard and brittle. Syd Lawton had a batch of around a dozen sent over from the factory and asked me if I would heat treat them, my reward being able to keep one of them for my bike. As he handed them over one fell onto the floor and shattered like glass. They were that hard. I was able to stress relieve the rods and temper them to the correct hardness, I seem to recollect that did the trick. A much better rod soon appeared with two machined ribs running round both the little and big eyes. The main body of the rod was also fully machined out to give the 'I' section. Another useful modification that I did was to fit needle roller bearings to the rocker arms and make the shafts stronger. I was the first to fit extra tubes to the rear frame running from just above the top of the suspension unit to the top frame tube across the loop towards the swinging arm pivot. This made the rear end much more rigid, When Syd saw this mod, he thought it was such a good idea that he told the factory and soon some of their bikes started appearing with the same addition.

The big highlight for me of the 1964 season was the Manx Grand Prix. This was the first year since 1948 that a 250cc event had been staged and then on the shorter Clypse circuit, the 1964 race was held over the full 37.75 mile Mountain Course. In practice, I had been getting around quite fast but had a serious engine problem and it looked as if I would be a non- starter but thanks to Syd Lawton who lent me his spare engine in the final practice period I was on top of the leader board with the fastest lap of all sessions.

In the race, newcomer Gordon Keith was flying and led the race, I was lying about fourth or fifth when on the first lap my rear brake torque arm broke and so I had no back brake for the rest of the race. The next thing to go was the rev counter drive so it was 'play it by ear', then the front fairing bracket broke and the wind was lifting the fairing way up high but seemed to be hanging on so I pressed on. Going over the Mountain for the third and final time I saw a massive crow in the road in front of me, it took off and flew straight into my screen smashing a large hole in it. Bouncing off the screen it hit my helmet with a thud and nearly knocked me off my bike! I carried on over the Mountain and down to Governor's Bridge. As I accelerated out of the dip I thought 'Good nearly home,' and as the revs rose there was a bang and the engine died completely, I whipped in the clutch and coasted silently towards the finishing line. The man holding the chequered flag was looking the other way and did not hear me coming I had to yell 'Hey, over here!' and so I finished third at 83.08mph average speed behind Gordon Keith (Greeves) and Rex Butcher (Aermacchi). The reason that my engine suddenly stopped was a broken contact breaker arm and I was very lucky that it had happened where it did and not up the mountain. Syd Lawton was delighted with the race results and I am sure that Rex was on one of his bikes. Syd was to come to my

1963 long-stroke fitted with short-stroke engine, big Oldani front brake and works megaphone, purchased at Mallory Park, in September 1965, from Renzo Pasolini.

rescue again on several other occasions later that year, one of which was at Snetterton when I had again blown my engine at the big International Meeting. He had some of the latest bikes from the factory, and one of them was identical to the works racers as ridden by Alberto Pagani and Gilberto Milani. This bike was given to me to ride and just after adjusting handlebar positions etc and getting the bike ready for the race, Dave Degens (Syd's rider at the time) suddenly wanted my bike and I had no other option but to let him have it. I was then given Dave's older model to ride. I shall never forget what he said when I asked him what to rev the engine to. He replied with a grin, 'Just rev it to the M on SMITHS!' This I did and the bike flew and we both finished well up and in the money. I had a very good start to the 1964 season with a win at the Spring National meeting at my favourite circuit, Scarborough. Again this was on Syd's bike, then the Southern 100 and later on at Scarborough again finishing third in the International behind Alan Shepherd on his works MZ and John Ashworth on a very fast Padgett Yamaha.

It was in 1965 at Mallory Park that I bought a 250 short-stroke engine, big Oldani front brake and long works megaphone from the famous Renzo Pasolini who was the factory rider for Aermacchi. This motor was very fast and reliable, but all these 250 four-strokes were beginning to be blown off by the new production racing Yamahas. I was approached by Peter Padgett from Batley to ride his Yamahas and was very pleased to do so. I decided to sell the Aermacchi in 1969, which I did for the princely sum of £200 to Geoff Monty. Previously, I had sold the old long-stroke engine all in pieces with damaged head etc. 12 years later in 1981 I had a phone call form Charles Smith, a collector from Crick.He asked me if I wanted to see the bike. I was not too bothered, but went to see it. It was still green, as it was when I sold it and in very good shape, having been owned and raced by Graham Thomas, a Rolls Royce engineer from Derby, until 1977. We put petrol in it and a good battery and I pulled it back on compression and three paces later it was running and sounded great... the old team back again. I was not really interested in getting it back but my wife convinced me that it could be a very good investment... how right she would be! My wife and I managed to raise the cash and I bought the bike back. After a full winter restoration I paraded this bike at most short circuits and it completed many TT laps of honour for the following 8 or 9 years. Domestic life and raising a family eventually took priority and the Aermacchi was used less and less. Meanwhile the old long-stroke engine was to reappear in another frame and was bought by David James in 1984. David contacted me 1985 and asked me to confirm the authenticity of the engine and frame, the engine was correct but the frame wasn't. He asked me if I ever wanted to sell the bike to him, he would be able to reunite the original engine and frame.

During the period that I was parading the Aermacchi, I was using a Volkswagen Camper to transport the bike and we used this van to go on camping and holiday trips. One dark night at around 4 am the van was stolen off our driveway and never seen again. It seemed that the time had come to contact David James and see if he still wanted to buy the bike and therefore reunite the original engine with its correct frame. David did indeed buy the bike from me in February 1993. He raced the Aermacchi with much success until 2003 when he became seriously ill and very sadly died in 2004. I believe David's family have kept the bike for the moment.

Terry Grotefeld

1964 Manx Grand Prix (4 laps-150.93 miles)

1st:	G. A. Keith (Greeves)	19th:	R. Waterer (Greeves)
2nd:	R. E. Butcher (Aermacchi)	20th:	R. Forrester (Aermacchi)
3rd:	T. P. Grotefeld (Aermacchi)	21st:	A. .H. Dyde (Royal Enfield)
4th:	J. Blanchard (Aermacchi)	22nd:	R. Kewley (Greeves)
5th:	G. Collins (Yamaha)	23rd:	R. A. Eldridge (Greeves)
6th:	G. A. Borland (Aermacchi)	24th:	W. Donnelly (Honda)
7th:	J. A. Wetherall (Cotton)	25th:	D. Padgett (Yamaha)
8th:	R. Ashcroft (Yamaha)	26th:	M. W. J. Meddings (Yamaha)
9th:	B. A. Ball (FB Mondial)	27th:	D. Rose (Honda)
10th:	R. F. Pladdys (Honda)	28th:	S. K. Suttcliffe (Aermacchi)
11th:	B. Scully (Greevesl)	29th:	I. E. Duncan (Aermacchi)
12th:	A. F. Dickinson (DMW)	30th:	B. J. Davis (Ducati)
13th:	C. H, Ward (LE/ BSA)	31st:	R. Spinks (Yamaha)
14th:	B. Lawton (Aermacchi)	32nd:	W. G. Beson (Honda)
15th:	K. G. P. Way (Cotton)	33rd:	B. R. Jefferies (Ducati)
16th:	T. G. Pallister (Cotton)	34th:	J. Shacklady (Greeves)
17th:	M. Chatterton (DMW)	35th:	P. Berwick (JD Dot)
18th:	C. V. Wallis (Cotton)	36th:	H. S. Woods (Bultaco)

The Terry Grotefeld machine as it was in the 1980s; still with 1961 seat, 1965 pre-production dry-clutch engine and exhaust, plus 230mm Oldani front brake, all ex-Pasolini.

6 Transatlantic Partnership

Riders Handbook for the 1961 Harley-Davidson Wisconsin (Ala Bianca on the Italian market).First model to be marketed in the States after the Harley-Davidson buy-in of 1960.

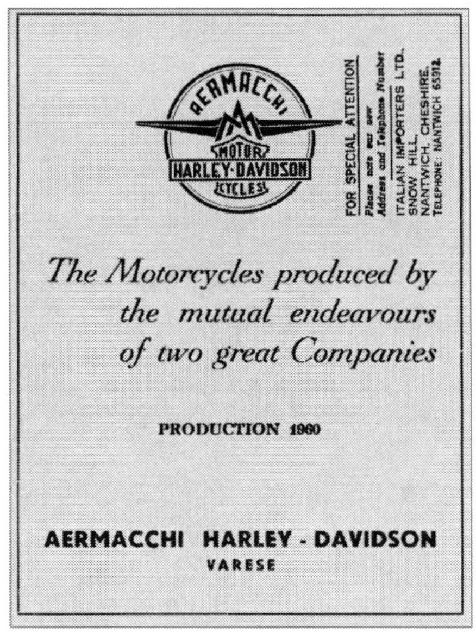

One of the strangest of all motorcycle marriages came about in 1960. It involved Italian Aermacchi and American Harley-Davidson partners. At first this seemed most unlikely, with Aermacchi having built its name upon small, lightweight machines, whilst Harley-Davidson was famous mainly for heavyweight, large capacity bikes.

It came about like this: In 1954 Harley produced 12,250 machines and by 1955 this was down to 9,550. At this time sales were booming. Worldwide motorcycle sales were on the up, but Harley's figures didn't improve during 1956, 57 or 58. So, the Harley-Davidson Development Committee held a crucial meeting on the 17th June 1959, and concluded that the best way of improving matters was to 'import a middleweight motorcycle'.

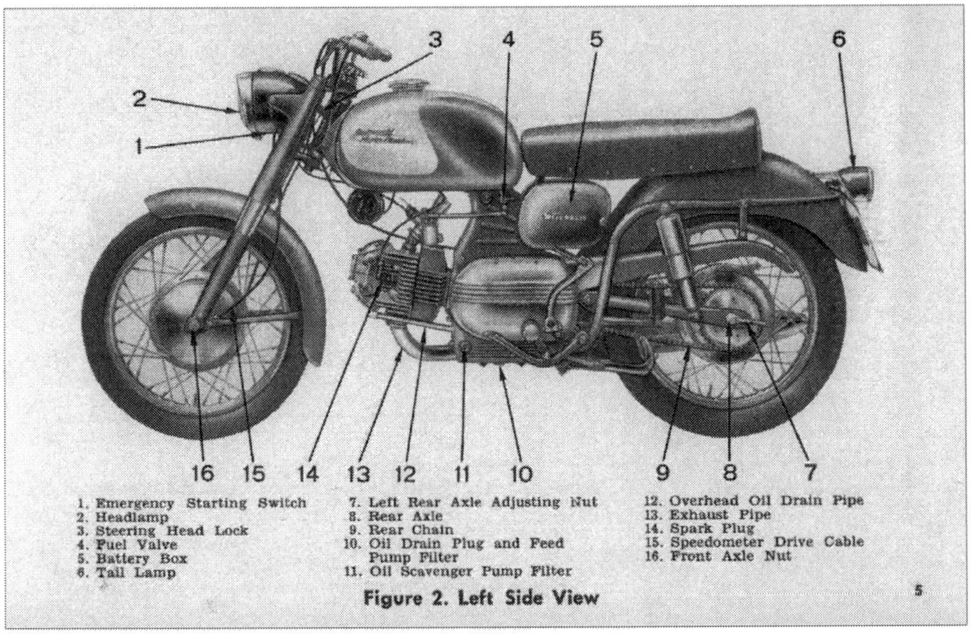

Owners' Handbook, illustrations, showing features of the Harley-Davidson 250cc Sprint C Wisconsin, 1961.

William H. Davidson and his wife travelled to several of Europe's leading motorcycle manufacturers. The one that impressed Davidson most, for a number of reasons, was Aeronautica Macchi of Varese, Italy. Why? Well, for one thing, matters would not be complicated by existing agreements with American firms, which would have been the case for the majority of British marques; BSA, Triumph, Norton, AJS, Matchless and the like. In any case, most of these companies concentrated their efforts in the larger capacity classes. The German industry had by this time all but collapsed. This left the Italians. Ducati, Parilla, Benelli and Gilera were already represented in the States. It was felt that Moto Guzzi was unsuitable, because its range consisted of long-running, old-fashioned big singles. There was of course a host of smaller marques, but these didn't have either the track record or the production capacity required by the Americans.

Macchi was a long-established engineering company which was well-respected, particularly for aircraft. And it was also the 250cc ohv horizontal single that impressed the Harley-Davidson committee following Davidson's subsequent report.

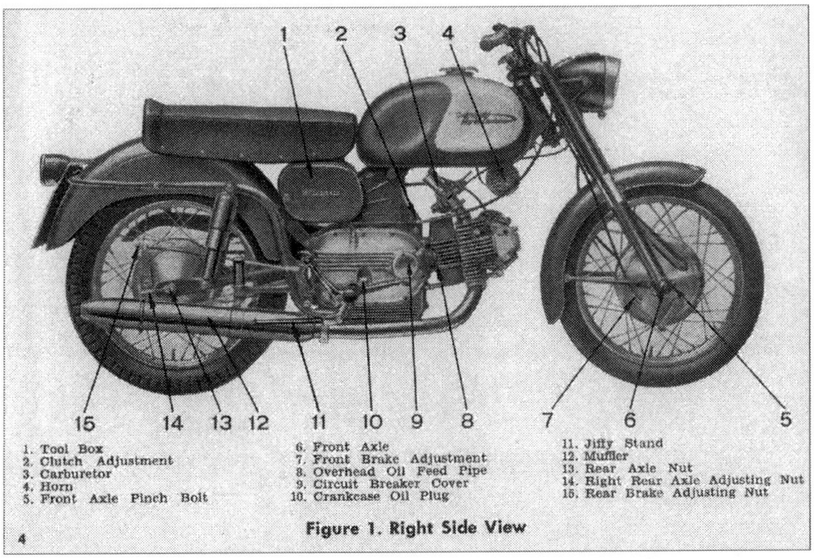

1962 Sprint C

Wisconsin

Engine:	Air-cooled ohv single, horizontal cylinder, alloy head, cast iron barrel, built up crankshaft with roller bearing big-end, vertically split aluminium crankcases, unit construction
Bore:	66mm
Stroke:	72mm
Displacement:	246.2 cc
Comp ratio:	8.5:1
Lubrication:	Gear type pump; wet sump
Ignition:	6 volt battery/coil; crankshaft mounted generator
Carburettor:	Dell'Orto
Primary drive:	Helical gears; 26 x 65 tooth
Final drive:	Chain, 16 tooth gearbox sprocket, 38 tooth rear wheel sprocket
Gearbox:	4-speed, foot change
Frame:	Spine type
Front suspension:	Enclosed telescopic
Rear suspension:	Swinging arm, twin shocks
Front brake:	Full width, alloy drum, SLS
Rear brake:	Full width, alloy drum, SLS
Tyres:	Front 3.00 x 17, rear 3.00 x 17

General Specifications

Wheel base:	52 in (1320mm)
Ground clearance:	5 in (127mm)
Seat height:	29 in (736mm)
Fuel capacity:	4 gal (18ltr)
Dry weight:	261lb (118kg)
Maximum power:	16bhp@6,500rpm
1963 on:	(18bhp@6,700rpm)
Top speed:	82mph (9140kph)

This was to result in an agreement whereby the motorcycle division of Aeronautica Macchi sold exactly half its share capital to the Americans for $247,209. David K Wright, author of 'The Harley-Davidson Motor Company', an official eighty-year history published in 1983 called this 'a bargain, even in those days.' The actual agreement was signed in April 1960, with the first motorcycle delivery being made in September 1960, well in time for 1961 season. And the Harley-Davidson management decided to send its chief engineer, Wilbur Petri to the Varese factory. Initially Petri was promised that his stay would be 'a maximum of 2 years', but in fact he stayed until 1967!

The first Stateside Italian import was the Sprint C, named the Wisconsin, after Harley-Davidson's home state. Essentially this was the European Ala Bianco 250cc touring bike (see Chapter 9) itself based closely on the Ala Verde sportster and Ala d'Oro road racer.

1963 Sprint H

Engine:	Air-cooled ohv single, horizontal cylinder, alloy head, cast iron barrel, built-up crankshaft with roller bearing big-end, vertically split aluminium crankcases, unit construction
Bore:	66mm
Stroke:	72mm
Displacement:	242.2 cc
Comp ratio:	9.2:1
Lubrication:	Gear type pump; wet sump
Ignition:	6 volt battery/coil; crankshaft mounted generator
Carburettor:	Dell'Orto
Primary drive:	Helical gears; 26 x 65 tooth
Final drive:	Chain, 15 tooth gearbox, 38 tooth rear wheel sprocket
Gearbox:	4-speed, foot change
Frame:	Spine type
Front suspension:	Telescopic oil damped fork
Rear suspension:	Swinging arm, twin shocks
Front Brake:	Full width, alloy drum, SLS
Rear Brake:	Full width, alloy drum, SLS
Tyres:	Front 3.00 x 18; rear 3.50 x 18

General Specifications

Wheel base:	53.5 in (1358mm)
Ground clearance:	5.3 in (134mm)
Seat height:	31 in (790mm)
Fuel capacity:	3 gal (13.5ltr)
Dry weight:	271lb (123kg)
Maximum power:	21bhp@6,700rpm
Top speed:	78mph (130kph)

In fact, in the first year of trading, this was the only Aermacchi model to be offered to American customers. Generally it was well received and sold well. One has to remember that, at the time, the Japanese were only just beginning their export drive into America, and that in performance and styling terms the Sprint C Wisconsin was a class act. Unfortunately, whereas Honda progressed, the Sprint C, and the other models which followed from Varese didn't. Weaknesses such as 6 volt lighting and poor quality switchgear, wiring, paint and chrome work didn't help. The Honda CB72 and CB77 (250 and 305 cc) were not only cheaper, but also had twin cylinders, 12 volt electrics, decent wiring and were generally more sophisticated, with push button starting and direction indicators for example.

But even with the faults mentioned above the early Sprint was well liked by both dealers and customers during the years 1961-1965.

For the 1963 model year the new Sprint H was introduced. This was what is best described as a Street Scrambler or Green Lanes Bike. Not a serious competition model, but the ordinary road rider could take it off road for enjoyment and seeing the countryside. For the pukka dirt bike rider there was the Scrambler which, stripped of the normal road equipment, was particularly successful in quarter mile short track events. This was coded CRTT or CRS (TT dirt track, S Scrambler).
By 1964 the Aermacchi range in the USA comprised the Sprint C, Sprint H and Motocross (Scrambler).

When this photograph was taken in spring 1962, the American market was already becoming a major sales outlet for the Varese factory. The 1961 250cc World Champion Mike Hailwood tries out the 1962 250cc Sprint H. With Mike is Nello Pagani, father of Aermacchi rider Alberto Pagani.

250cc Sprint from the Harley-Davidson factory brochure of 1963.

Then for 1965 the Americans decided to re-import the existing Zeffretto, a step thru moped and, from 1966 the Zeffiro Sport, an ultra lightweight sports style motorcycle. Both were powered essentially by the same 49.66cc (38.8 x 42mm) single cylinder two-stroke engine with piston port induction. These were designated M (Moped) and MS (Sport). One failing, at least on the MS sports motorcycle, was that the engine had only three gears with internal rations of 1st 3.18:1, 2nd 1.78 and 3rd 1.17.

Although widely advertised, neither of the miniature two-strokes were very successful in the States. In fact, some 4,000 M50 mopeds had to be savagely discounted in 1966, following almost zero sales. And the introduction of the new-for-that-year M65 and M65 Sport didn't help.

The 65cc engined machines continued to be offered until as late as 1972, but by then as the Shortster, a Monkey bike machine with 10 inch wheels and fold down handlebars, the forerunner to the similar X90 of 1973-74.

For the 1967 model year, the 250cc ohv horizontal engine was redesigned, with new short-stroke bore and stroke dimensions of 72 x 61mm, giving an exact

Left; Features of the 1963 Sprint which Harley-Davidson described as 'Race Bred with Lion Hearted Power!'
Right: American buyers were also offered a Street Scrambler version of the 1963 Sprint model.

1964 Sprint H was the 'sports' model with smaller tank, no toolboxes, abreviated front mudguard and 25bhp. Cycle World magazine achieved a maximum speed of 76mph and 0 to 60 in 15 seconds.

engine displacement of 248.4cc. An aluminium cylinder barrel replaced the former cast iron component, but the Americans still only got a 4-speed gearbox. The sprint H title continued as did the prefixes CRTT and CRS, but the Sprint C was axed and in its place came the SS (Super Sport). No less than 7,000 of the SS machines would come into the States that year, compared to a mere 600 Sprint Cs in 1966. Conversely only 2,000 Sprint H models were sold in 1967, compared to 4,700 the previous year. Also the American version of the new short-stroke 250cc had several styling differences to the European versions of the time; a one piece rocker cover, the Speedo and tacho mounted in a single aluminium panel and different tank and exhaust, etc.

1965 Sprint C Touring.

Next, in 1968 came ML125 Rapido, based on the new-for-1967 European Alletta. This was a 123.5cc piston port two-stroke single, but this time a with a 4-speed 'box. The 56 x 51mm bore and stroke dimensions were short-stroke in keeping with the new fashion. For the full specification readers are advised to consult Chapter 9.

In 1970 the MLS125 arrived to replace the ML, but was very similar. An identifying feature of the series was the front fork. 1969 and earlier used metal spring covers for the front forks, 1970-

1971 rubber gaiters, and 1972 onwards exposed stanchions with internal springs. 125cc developments continued with the TX in 1973, SX in 1974 and SXT in 1975 onwards. All three of these employed a Japanese Mikuni oil pump; whereas the ML and MLS were Petroil. Another change at that time was from 6 to 12 volts. The TX, SX and SXT also came into the post Aermacchi era and were in fact marketed as AMF Harley-Davidsons (see Chapter 12). It has to be said that in the 21st century none of the various two-stroke Aermacchi/ Harley-Davidson models are collected or restored; except for the limited production RR250 and 350 road racers.

Not so the various Sprint models. And the final incarnation of the theme was the 350cc. The 344cc (74 x 80mm) engine was, like the European version, a long stroke type, which itself shared the same size and dimensions as the customer Ala d'Oro road racer. However, unlike the racer, the roadster had only 4-speeds (later rectified on the European TV in 1971, and other models from the 1973 model year.) And of course the valve gear was still push rod operated. The 1969 Harley-Davidson Sprint SS was very similar to the European GTS 350 (see Chapter 9). Both featured a single exhaust port with branched header pipe and two silencers, one either side of the machine.

In a 1970 road test 'Cycle' reported, ' You would never mistake the Italian built Harley-Davidson Sprint for a Japanese machine, it goes its own definite way in styling. The gas tank and fenders of our test bike were painted a handsome, grassy shade of candy apple green. The simple tank is unmarred by anything other than the appliqué brand names, and the shape of the tank fits the

1965 Sprint H.

1967 Sprint H with low-level exhaust, chrome air filter container and fork gaiters.

personality of the bike perfectly. The tank reaches right out and grabs you, too, because the headlong is black and the toolbox is matt black, so there's nothing to distract your eye.'

However, 'Cycle' had a few compaints too. 'You can feel the vibration grow as the load increases and you open the throttle. On the freeway you can barely hold your feet on the pegs and the instruments are gyrating so that you can't read them accurately. Finally you pull off onto the country road and through the puddle and you drench yourself and get a face shield full of mud. You stop to dry off and you realise there isn't a centre stand, so you kick down the side stand.'

However the lack of noise got full praise; 'The only sound from the engine is an exhaust tappet click- the two mufflers that come from one header pipe made hardly any noise.'

The 1971 'Cycle Buyers Guide' tested six 350 bikes. These were the RD 350 Yamaha, CB 350 Honda, A7 Kawasaki, T 350 Suzuki, GTR Bridgestone and the Sprint SS. The fastest over the quarter mile was the Kawasaki in 14.35 seconds, a speed of 91.55mph; the slowest was the Sprint in 15.85 seconds; 80.64mph. However, the Sprint came out on top in fuel consumption, achieving 58.3 mpg

1969 Sprint H 344cc with much larger 'streamlined' valve gear cover.

In Italy and the rest of Europe Aermacchi's 47.61cc (38 x 42mm) 3-speed 2-stroke was known as Zeffiretto. In the States this commuter moped was listed as the M50. First sold in 1963 it was never a big seller, Harley-Davidson never really embracing flyweight machines.

(remember an American gallon is smaller than an imperial unit!) whereas the Japanese two-strokes were hard pressed to reach 30mpg. Weight-wise, the Sprint was 308lb compared with the only other four-stroke, the Honda CB 350, at 344lb. The comparison test ended by saying, 'The Sprint is refined way past its design capabilities, boasting good detailing and good hardware. But it needs a two cylinder engine and it needs a 5-speed gearbox and it needs, after all is said and done, to be made into a trail bike or a scrambler. It's street competition has simply gotten too good.'

Harley-Davidson

Sprint Production Figures

Displacement:
246.2cc, 66 x 72mm, 18bhp@6,700rpm

Year	Model	Production
1961	Model C	N/A
Model 61 CRTT Off-road		N/A
1962	Model C	N/A
	Sprint H	N/A
1963	Sprint C	150
	Sprint H	1416
	63 CRTT	N/A
1964	Sprint C	230
	Sprint H	1550
	CR	50
1965	Sprint C	500
	Sprint H	2500
	CR	35
	CRS	175
1966	Sprint C	600
	Sprint H	4700
	CR	50
	CRS	350

Displacement:
248.4cc, 72 x 61mm, 21bhp@7,250rpm

Year	Model	Production
1967	Sprint H	2000
	SS	7000
	CRTT	35
	CRS	50
1968	Sprint H	
	SS	4150
	CRS	125

Displacement:
344cc, 74 x 80mm, 1969-72 25bhp@7,000rpm
1973-74 27bhp@7,000rpm

Year	Model	Production
1969	SS350	4575
	ERS	250
1970	SS350	4513
	ERS	102
1971	SS350	1500
	SX350	3920
	ERS	50
1972	SS350	3775
	SX350	2525
	ERS	50
1973	SS350	4137
	SX350	2431
1974	SS350	2500
	SX350	2085

Harley-Davidson did respond by introducing a 5-speed gearbox from 1973 model year in States and by the introduction of a trail bike version, the SX 350, from late 1971. However, it didn't go for a twin cylinder format, which would of course, have meant a far bigger cash outlay.

The European Roadster is discussed in Chapter 9. The development of the 1972 model year machines is charted in Chapter 12, which includes both the SX 350 and final variants of the SS 350.

On the sporting front in the States, one man stood out for his vital development work and feedback to the Varese factory. This was Dick O'Brien, the Harley-Davidson racing chief. It was largely O'Brien, together with his British counterpart Syd Lawton, who pushed the Italians into making the necessary changes to the Al d'Oro racer, which transformed them from being fast but fragile in the early days, to even faster but virtually bullet-proof by the time production ceased at the end of the 1960s.

1972 AMF Harley-Davidson 4-gear, Sprint SS (European models had 5). No tachometer or rubber-mounted handlebars. Small 'peanut' tank. Note dual exhaust header pipe and two silencers.

7 Off-Road Competition

Dutch rider Piét Kramer putting his Aermacchi moto-crosser through its paces.

The Aermacchi name is usually associated with road racing. However, the Varese company was also involved in off-road sport, most notably long distance trials and scrambling (moto-cross).

The 25th September 1952 issue of 'The MotorCycle' published a report of that year's International Six Days Trial. 'Among the unusual machines seen on Wednesday were the 125cc Aermacchis from Italy. Aermacchi machines have been displayed at the Milan and Geneva exhibitions, but this was the first time they appeared in the ISDT. Features include a leading link front fork and pivoted-fork rear springing. The rear fork and engine pivot together.' The trial was centred around the spa resort of Bad Aussee in Austria, 1,700 metres (5,500 ft) above sea level and 45 road miles east of Salzburg. In this instance none of the Aermacchis retired but neither did they win any medals. However, they did prove their reliability under the most demanding conditions.

Another facet of Aermacchi's involvement with motorcycle sport was motocross. This began in 1959 when the factory's race shop produced an experimental 500 class dirt racer. This was first seen in public at an Italian motocross meeting. It was ridden by the Italian born Belgian scrambler, Fernand Neri. The half-litre machine was dominated by its engine, which had a vertical cylinder and a massive but very light cambox. The twin camshafts were driven by a train of gears on the nearside of the motor, while the outside oil-pump, very strange on an off-road bike, was mounted at the base of the cam drive gear housing. Immediately above the oil pump was the contact breaker assembly.

Unlike so many of its Italian contemporaries, the 500 Aermacchi single had dry, rather than wet sump lubrication; the oil being carried in a separate tank under the saddle. Again, the exposed valve springs of hairpin type might have seemed out of place on a dirt racer, but this system had seldom led to complications on either of the Belgian FN machines or on Les Archer's famous ohc Norton motocrosser.

Originally, it had been intended to pipe the exhaust gases out under the footrest on the right-hand side of the machine but instead the pipe was on the left. Bore and stroke was 88 x 82mm and engine capacity was 498.48cc. The factory stated that on a compression ratio of 9:1, the motor produced 40bhp at the rear wheel, with the engine spinning at 6,200rpm.

During th early 1950s Aermacchi entered several teams in the ISDT, gaining vital experience in such events - and benefiting the series production models.

Aermacchi also built this experimental 498.48cc (88 x 82mm) dohc engined moto-crosser which was extensively tested in 1959 and 1960. Although it was powerful, it never entered production.

By 1960, the prototype 500 had its frame modified to follow 'Featherbed' Norton practice, with double front down-tubes replacing the original single unit. A cable-operator rear brake had also been modified to rod operation. Other details of interest included large diameter full-width brake hubs, 21 inch front wheel, a dry weight of 152kg (335lb) and the choice of either four or six-speed gearbox.

The 246cc Aermacchi with which works rider Lanfranco Angelini finished 3rd in the Italian Moto-Cross GP in June 1960.

Plans for production were never fulfilled, probably because of the success gained by Neri in the 1960 European Championship on a motocross version of the 246cc pushrod single. Instead, the factory built a production 250 moto-crosser which went on to win the Italian title five times in the hands of Lanfranco Angelini, and was also exported to North America for both off-road racing and flat-track events well into the 1960s.

The motocross development was also responsible for the first 350cc class Aermacchi flat single engine, with a 293cc (72 x 72mm) dirt model, which was constructed in 1961.

During the early 1960s the production 250cc moto-crosser was quite competitive, but as the decade wore on, a new breed of two-stroke appeared, bikes like the Greeves Challenger, Bultaco-Metisse and twin port CZ. But during the 1965 season New England Aermacchi-Harley-Davidson dealer Peter Preble and scrambler Bob Hicks joined forces to build an Aermacchi with which to take on the two-strokes.

Hicks had raced a Bultaco with some considerable success but was 'always on the lookout for something different.' At first, the Macchi was used for Saturday night's quarter mile dirt track events, where he wound up second in the Experts standings at the end of the season. He recalled later, 'We were familiar with the power output of the horizontal pushrod engine, which was considerable, and decided to see what could be done to improve the poor handling in the rough for it was pretty bad news when the going got bumpy.'

The long-stroke, wet-clutch engine had been modified. The compression ratio had been increased to 12:1 by shortening the cylinder. A larger, 30mm carb had been fitted, with the inlet port bored out to suit. There was also an oversize inlet

1971 125 R/C

Engine:	Air- cooled piston-port single cylinder two-stroke alloy head, cast- iron barrel, built-up crankshaft with roller bearing big-end, vertically-split aluminium crankcases, unit construction.
Bore:	56mm
Stroke:	50mm
Displacement:	123.15 cc
Comp ratio:	10.3:1
Lubrication:	Petroil mix
Ignition:	6 volt 60 watt flywheel magneto
Carburettor:	Dell 'Orto VHB 27A
Primary drive:	Helical gears
Final drive:	Chain, 12 tooth gearbox sprocket, 58 tooth rear wheel sprocket
Gearbox:	5-speeds; foot change
Frame:	Duplex cradle, all-steel construction
Front suspension:	Ceriani exposed stanchion telescopic fork
Rear suspension:	Swinging arm, twin shock absorbers
Front brake:	Drum SLS 135 x 25mm full width alloy hub
Rear brake:	Drum SLS 135 x 25mm full width alloy hub
Tyres:	Front 2.50 x 21, rear 3.50 x 18

General specifications

Wheel base:	51.2 inch (1300mm)
Seat height:	32 inch (815mm)
Fuel capacity:	2.64 gallon (12 litre)
Dry weight:	236lb (107kg)
Max power:	15.5bhp@8000rpm
Top speed:	75mph (125kph)

The SR100 had a string of victories at American events in the early 1970s, including the Baja 500 and Mexican 1000.

Harley-Davidson, as opposed to the earlier Aermacchi efforts, first Moto Cross GP was at the 1976 US round, when works rider Rex 'Rocket' Staten debuted this experimental 360cc machine.

valve, with the piston modified to accommodate it. The original sports profile camshaft had been replaced with a pukka dirt track racing component. These changes resulted in 'an engine that pulled beautifully from 6,000rpm to 10,000rpm; and with shimmed valve springs we had no float problems with the valves, nor did they strike pistons. This engine hauled the 245lb bike off the line and down the straights faster than the lighter 210lb Metisse could move, despite its claimed 35bhp.' It should also be remembered that the American specification was fitted with a semi-road race cylinder, 12:1 head of conventional fin pattern, and a different exhaust chamber.

The first modification Preble and Hicks made to the Aermacchi to suit it for scrambling was to replace the front forks with Bultaco components. However, Bob Hicks explained, 'After a successful debut, on a rather flat and hard track, winning the Experts 250 feature race, I found the bike still a bit weird handling.' The new forks were too long, so these were cut down by an inch. Then to bring what was described as the 'bouncy and unpredictable rear' into line, further modifications were carried out.

The swinging arm was lengthened by 2 inches, making the wheelbase a nominal 56 inches, whilst the Italian shocks were replaced by Jawa Motocross ones. Since these had 3½ inches of travel, and were 1½ inches longer than the stock Italian units, this pushed the swinging arm down to an undesirable angle for chain tension control. So, the top rear frame was cut off and replaced to re-locate the rear shock top mounts, about an inch higher and an inch further back.

The result was that steering was very precise over the rough, even in one very wet mud-bath where deep cement-type mud, and thin greasy mud on hard-pan made a vicious combination.

The only problem with the engine was continued ring breakage, a known problem on the long-stroke. The 10,000 plus revs attained by the long-stroke motor were just breaking the rings into little pieces and, in the hurly-burly of scrambling, it was impossible to keep the rpm under control most of the time. Even with compression rings breaking up, the motor would still reach 9,000rpm and match the Bultacos on speed, but the edge in power was then missing.

Eventually, the 'bumps and jumps' took their toll on the spine frame, with the frame bending by some two inches. This resulted in the frame having to have its correct wheelbase restored by means of 'hydraulic jacks.' It was then built-up, with a 'sub-frame of 1 inch water pipe to brace the spine against further bending.' This modification proved a total success.

In a total of 32 meetings (20 motocross and 12 dirt track) the team had to fit 10 sets of new rings. They then planned to modify the piston to take Dykes type rings, in the hope of reducing this trouble.

In contrast to the ring glitches, and the fact that the revs needed to be kept up, the gearbox, according to Hicks, proved 'fantastically slick, shifting without the clutch' and never needed attention.

As the time passed, the four-stroke was replaced by the two-stroke in dirt bike sport. And as with its road racers, Aermacchi responded by developing its own two-stroke dirt bikes. The first to arrive in 1971 was the 125 R/C, (Regolarita/Cross). This was powered by a tuned version of the existing Rapido roadster engine, which was also used as a base for the 125 Ala d'Oro road racer. This meant a 123.15cc (56 x 50mm) single with piston-port induction and vertical cylinder. Running on a compression ratio of 10.3:1, the R/C produced 15.5bhp at 8000rpm. It employed a square slide VHB 27mm Dell 'Orto carburettor and petroil mixture.

The frame was a particularly neat duplex cradle affair and, as was traditional with Aermacchi machines, Ceriani suspension both front and rear.

Other two-stroke off-road machines were developed (which are described in Chapter 12), but were produced exclusively for the American market. The 125 R/C bucked this trend, being a Europe-only product. But it had a relatively short production life and was discontinued after Harley-Davidson gained full control of the Varese plant in 1972.

Prototype MX250 Moto-Crosser built in mid-1970s used an engine based on an AMF Harley-Davidson SX250 trail bike, but with new top end, disc rear brake and highly unconventional rear suspension.

8 Dry-Clutch Racers

By 1966 the 250 Ala d'Oro had been given a new look and a shorter stroke, 72 x 61mm bore and stroke measurements, a dry-clutch had also been adopted.

Works development 344cc (74 x 80mm) machine as raced by Dave Degens in 1964 for Syd Lawton, seen here rounding Devils Elbow at Mallory Park. The first 350cc class engine from Aermacchi was an experimental 298cc which had its origins in moto-cross and had debuted in 1961.

As related in Chapter 5, Aermacchi began race testing the first dry-clutch models on their works development machines in late 1964. The majority of the 250 Ala d'Oro racers were wet-clutch bikes, things were different with the dry-clutch series.

One of the very first 'production' models to arrive in the UK was brought in by the British importer, Syd Lawton, for his rider Dave Degens. It was tested in May 1966 by Bruce Main-Smith for 'MotorCycling'. Compared with the Ala d'Oro ridden by Degens in 1965 (a 1964 wet-clutch version), the newer bike had better acceleration, a higher maximum speed and better handling.

The dry-clutch bike had been purchased by Syd second-hand from an Italian dealer in Varese and was alleged to have been the actual model to have won the Monza 6-hour race in May 1965, averaging 98mph.

On the Lawton machine, a 33mm exhaust valve had been fitted in place of a 31mm component, itself 2mm larger than the original wet-clutch 350 valve. Unfortunately the piston cutaway had not been enlarged. As the 'MotorCycling' test of May 1966 revealed, 'This, with a tendency to over-run fourth gear, resulted in the exhaust valve touching the piston during warm-up on Racer Test day.'

New for the USA market in 1966, Harley-Davidson CRTT 250cc engine, with magneto ignition (only on the 250 due to fixed advance making the 350 too dificult to start). The 250 engine now put out 33bhp, the 350 35bhp.

Already having connections with Syd Lawton (as co-rider with Phil Read on the Lawton 650SS Norton in the Thruxton 500 mile races), Brian Setchell rode this 250 Ala d'Oro in the Lightweight TT. This is the latest model that year with dry-clutch and short-stroke engine.

Bruce Main-Smith reported, 'Lawton removed the valve, straightened it with a hammer, and ground it in again. He also scraped the valve pocket and adjusted the over-run stops on the gear-change.' And not only did the engine perform well for the full test session, but it carried Degens through a hectic Easter programme without any further trouble.

The bike, as purchased and tested had 35mm Ceriani forks, full fairing and small Oldani brakes. It was delivered with Ceriani rear suspension units, but these were changed for British Girlings.

Ridden by Dave Degens for Syd Lawton, this machine has a 1964/65 type frame, with earlier tank and later 35mm Ceriani forks.

Swiss rider/engineer Othmar Drixl at Brands Hatch with his 250 Aermacchi leads Len Rodda on a similar bike and Honda rider Ron Pladdys.

Monza, Sunday 14th September 1966. Factory riders Renzo Pasolini (left, without helmet) and Alberto Pagani. The pair had just finished a brilliant 2nd and 3rd in the 350cc Italian Grand Prix.

When interviewed regarding the preparation of the machine Syd said, 'Although it's probably not strictly necessary, I change the oil after every meeting. After every alternate outing, I check the valve clearances and the contact breaker gap. But the new-type Dansi points on this engine seem very reliable and rarely need re-setting. The timing has varied 1° in the 250 racing miles since purchase. Nor have the valve clearances lifted out by more than one thou. The rear chain, which is washed and immersed in hot 'Linklyfe' every 100 miles, lasts a whole season. The front forks need no attention, they are filled with SAE5 on the Aermacchi, and the rear end swinging arm pivot requires just a little grease every 150 to 200 miles. The rest of the bike is equally trouble-free, except for one item, if the rider

For the 1967 season a switch was made from Oldani to Ceriani brakes, a four leading shoe 230mm at the front... ...and twin leading shoe 200mm at the rear.

Dave Degens with the Lawton works 350, Brands Hatch, 1st May 1966.

Australian Kel Curruther's 350 Aermacchi Metisse leading Honda-mounted Mike Hailwood during the Race of Aces, Snetterton, Summer 1967.

leans the model to the left for long on the grid, the inlet tract fills up with petrol. Lodge RL50 or 51 plugs are entirely suitable. Cold start, warm up and race on the same plug. The surprising thing about this engine is the quite ordinary workmanship. No one has lightened the piston or the valve gear. Even the rockers retain their manufacturing marks and sharp edges. A lot of reciprocating weight could be saved and reliability further enhanced by grinding away the surplus metal on the rockers and polishing them for fatigue resistance.'

As for the test itself Bruce Main-Smith commented, 'This (the Aermacchi) seemed to require at least 500rpm more than a 350 Manx.' But it ran on correspondingly higher at the top end. Also, 'The power band and the gear ratios were such that there was no trouble at all in keeping clear of megaphonitis. The motor was smooth and free from any fuss...The 5-speed box, a good one to use, suited the horizontal single very well indeed. I had exactly the cog necessary for each corner.'

Dave Degens, 250 Ala d'Oro, Mallory Park 29th October 1967. Note Oldani front brake - preferred by Degens due to its adequate stopping power and light weight.

A trio of Aermacchis lead the field at the start of the 250cc race, Mallory Park on 17th September 1967 - Dave Degens (8), Paul Smart (40) and Ray Pickrell (28).

A 1968 248cc (72 x 61mm) Ala d'Oro racing engine with dry-clutch 5-speeds, forged piston, short con-rod, 'H' cam. 10mm diameter chrome tappets were standard, but later 12mm components gave greater reliability.

Handling was described by Main-Smith as 'excellent, despite the rear units "pattering" over a travel of $^{3}/_{4}$ inch on some sections of the circuit, especially when accelerating very hard in third coming out Copse.' However, 'The front forks were beyond criticism. Springing was not hard yet there was no disconcerting dipping during heavy front-wheel braking.' Dave Degens found that after the rear shocks had been changed for the Girlings, the Aermacchi handled as well as a good 350cc Norton. Bruce Main-Smith went even further, 'For my money, I prefer it. The bike gave me the sensation of being a lot lighter and higher than a Manx and considerably more throwable on sharp bends. Yet I had no worries about straight-ahead navigation as can sometimes come with a shortish wheelbase.'

Bruce Main-Smith's lap times were a personal 350 best in this Racer Test series, despite under-revving because the rev counter drive gear-box had failed.'

The 1969 works 350cc Aermacchi with 38mm Dell'Orto carburettor. To clear the frame the carb was slightly inclined to the left, this required a modified tank as shown, normally the hole was covered with gauze.

Alan Barnett (Lawton Aermacchi) at Thruxton, 12th April 1970.

1967, 344cc Ala d'Oro. Note the non-standard cutaway clutch cover modification, also original SAFA battery.

1968 250/350 Ala d'Oro

350 in brackets where different

Engine:	Air-cooled ohv single with horizontal cylinder alloy head and barrel, forged 3-ring piston, built-up crankshaft with double roller bearing big-end, vertically split aluminium crankcases, unit construction, hand-forged chrome nickel moly con-rod; outside flywheel
Bore:	72mm (74mm)
Stroke:	61mm (80mm)
Displacement:	248.3cc (344cc)
Comp ratio:	11.5:1 (11.1)
Lubrication:	Only one oil pump, feed pump, and greater use of gravity scavenge return system
Carburettor:	Dell'Orto SS1 racing type
Primary drive:	Gears; dry-clutch
Final drive:	Chain
Gearbox:	5-speed, foot-change, close ratio, re-designed
Frame:	Spine type
Front suspension:	Ceriani GP racing telescopic, oil-damped forks
Rear suspension:	Swinging arm, twin shock absorbers
Front brake:	Ceriani 230mm 4LS double-sided drum
Rear brake:	Ceriani 200mm 2LS
Tyres:	18 inch front and rear

General specifications

Ground clearance:	6 inch, 152mm
Seat height:	29 inch, 736mm
Fuel capacity:	3 gallon, 13.5 litre
Dry weight:	210lb, 95kg
Maximum power:	30bhp@10,000rpm (40bhp@8,400rpm)
Top speed:	120mph, 193kph (130mph, 209kph)

Factory fresh 1968 Ala d'Oro. By now these over-the-counter racers were extremely reliable - which is more than could be said for the earlier bikes.

Tommy Robb, right, seated on a Bultaco, raced a 350 Aermacchi in 1968.He is seen here with his 'crew' at Alicante, Spain, February that year.

Another Irish rider to campaign Aermacchi machinery with considerable success was Brian Steenson, here with sponsor Ronnie Conn, 1968.

The 1.6 mile Silverstone Club circuit was used, and it was a cold spring day. 'Light wind generally half-front along timing straight and assisting from Copse to Maggot's; little effect elsewhere. Dry track.'

The track-speed equivalent to peak power (7,800rpm) would have been 118mph. However, the best actual speeds achieved were Degen's 112.5mph, with Main-Smith at 107.14mph.

Besides the introduction of a dry-clutch for 1965, there was also a reinforced frame and a new oil pump. Speaking of the oil pump in March 1965, Syd Lawton announced, 'We had some trouble with them last year and I advise anyone who's got a 1964 Aermacchi to contact me for a replacement. I'll be getting a supply of the redesigned pumps soon.'

For the 1966 season the 250 (now capable of 9,800rpm), by now also with dry-clutch, received wider gears, and a number of engine improvements listed in the 'Year by Year Changes' section within this chapter. The 1966 350 was very much as per the smaller model except for connecting-rod, larger carb, larger diameter flywheels, longer holding down studs and pushrods, and 8,000rpm.

For 1967 there was a new combustion chamber shape, retaining the previous squish design. The ports were also altered. Float-bowl size was increased from the SS2 type to the SS1. The compression ratio was increased, with the ignition timing retarded to suit. Aermacchi claimed an additional 5bhp on the 350 making

Tommy Robb, with the near stock 350 Aermacchi he raced in 1968.

Brian Steenson winning at Carrowdore, September 1968.

Brian Steenson restarts his Ronnie Conn Aermacchi after falling during the North West 200, 24th May, 1969.

1969 'Long-stroke' production model (344cc).

40 in total. There were also smaller internal flywheels, a large external flywheel and three main bearings instead of two. Camshaft support was also improved, but the cam profile remained unchanged. Some engines had closer ratio 'A' gears; the others having wider B ratios.

It should also be noted that factory 250 'specials' and genuine works engines had special eccentric rockers; this also applying later to the short-stroke 350, 382 and 402cc works engines. Full changes are catalogued separately in the boxed section within this chapter.

Austrian GP, 1st May 1969. Works development rider Gilberto Milani heads the field in the 350cc race.

'MotorCycling' carried out what was their third 350 Ala d'Oro track test for the 7th January 1967 issue. This time Bruce Main-Smith was joined by John Cooper to put the machine through its paces at Silverstone. This actual bike had been built privately by Alberto Pagani using a factory- supplied May 1966 engine, and itself virtually a pre-production prototype using possible 1967 model year features.

The frame was the 1967 type and had Fontana brakes, selected by Pagani in Italy in place of the stock Ceriani units, for their availability. The batch of eighteen 1967 350s, imported by Lawton and Wilson were supplied with Ceriani assemblies.

Bruce Main-Smith described the Fontana brakes fitted to the test machine as superlative. 'Much lighter than yesteryear, the air-cooled dry-clutch on this model was very sweet and Lawton claimed, completely abuse-proof. Certainly it ran straight through the acceleration-graph runs without protest... Handling has improved. Some of the gain is attributable to the stiffer frame, some to Lawton's lengthy and ceaseless labour over suspension units. The sensation is of sitting up high on a short-wheelbase bike.'

The 1970 Junior TT was the Aermacchi single's swansong in the Isle of Man TT. But what a swansong! Alan Barnett took the Lawton-entered bike to runner-up position behind World Champion Giacomo Agostini's MV. In the process he put in a fastest lap at a shade under 100mph.

Australian Jack Findlay heads for 3rd place in the 1969 Junior TT aboard Francis Beart's immaculate lightweight Aermacchi.

After looking set to clean up with runner-up spot behind Agostini's MV in the 1969 Junior TT, Australian Kel Carruthers suffered a flat battery on the last lap, with his works entered bike.

Bob Farmer taking his 250 Ala d'Oro to victory in the Lightweight Manx Grand Prix, September 1966.

Australian Ken Duperouzel in action at a classic race event 'down under' on his 1968 344cc Ala d'Oro.

Jim Porter at Snetterton, 13th May 1990. Jim won the CRMC 350cc championship title first on an ex-Mick Walker Saxon framed Ducati and later on an Aermacchi.

On the same 1.6 mile Silverstone Club circuit, both Cooper and Main-Smith were recorded through the speed trap at 115.36mph; Cooper lapping in 79.5mph, Main-Smith 76.5mph.

AT the end of the year Syd Lawton held what amounted to a couple of open days at Snetterton in Norfolk, Tuesday and Wednesday, 5th and 6th of December. 'MotorCycle' declared, 'Syd's confidence is justified, enthusiasm for one's own products is understandable. But complete trust and confidence in someone else's goods has been the undoing of many a good man, not least in the motorcycle

The ex-Ken Redfern 1969 Aermacchi at the Bob McIntyre Memorial races, Knockhill 14th, 15th June 1986, it was then owned by Vic Lerigo. In the background is a Royal Enfield GP5 two-stroke.

Dutch riders Theo Louwes (31) and Leo Bovee (39) with their 350 Ala d'Oros, Summer of 1970, battling it out at the Roable circuit.

Eric Cheers came out of retirement to race this immaculate 382cc Aermacchi during the early days of classic racing. This photograph was taken at Oulton Park in Summer 1985.

sphere. But not Syd Lawton, the sporting Southampton dealer who has been Aermacchi concessionaire for the past four years. He knows the Italian pushrod singles are reliable. And on Tuesday and Wednesday he proved it by allowing over 50 riders of varying ability to thrash three 250s and four 350s around Snetterton for a total of 2,250 miles.'

The fifty-five riders who turned up included Peter Butler, Rex Butcher, Richard Difazio, Tony Melody and Ray Pickrell. In addition, some 200 laps were completed on Tuesday by pressmen and Syd's own staff, who were busy sorting out the machines for the serious session the following day.

David Dixon tested a 250 at the Lawton giant testing session. 'Although the rev meter was blood-lined at 9,500rpm, Syd recommended letting the unit go on to 10,200 in the gears. This it did most willingly, it even buzzed up to 10,400 before power fell off, and its smoothness

The classic racing movement began in the late 1970s. Former racer Dave Bedlington's 1968 350 Aermacchi is seen at Cadwell Park in 1981.

was immediately impressive... Early Aermacchi 250s had a remarkably flat power curve like a roadster's. But in recent years, the curve has been narrowed to the point where megaphonitis now clears and usable urge starts at 7,000rpm.'

The closeness of the gears also meant that the 1967 250 could be 'kept on the boil quite readily, more easily than with, say, a two-stroke twin. Nor was the power delivery so sharp as on a stroker.' Interestingly, David Dixon also sampled what Syd Lawton called his 'best 250,' the one usually ridden by Dave Degens. Dixon noted, 'Its engine was noticeably faster, better acceleration, willing to buzz over 10,500rpm.' He was told that Dave takes it to 11,000!

An Aermacchi competitor at North Gloucestershire's club races, Wroughton, mid-1970s.

Ian Lawton, Aermacchi 250, leads a CRMC race at Donington Park, 1981.

When it came to trying the 350 model, a Drixton-framed bike, David Dixon had this to say, 'Rough compared with the smaller units, its engine developed markedly more punch lower down the scale. It would pull away from around 4,000rpm, with real punch coming in at 5,000 to 5,500rpm. The rush up to recommended peak of 8,000 in the gears was obviously not so rapid as on the 250. And, on the Norwich Straight I found it quicker to engage top at 7,500rpm.' And what problems were there over the two days? The answer is not many, three accidents, none serious, a few flat batteries and a blown pushrod tunnel oil seal on a 250.

The following year saw what was, in effect, the final development of the Aermacchi ohv horizontal single, at least regarding the standard production racers. And most details were relatively minor. However, there was a surprising number of what are best described as 'works specials'. These included a 349cc (77 x 75mm), 348cc (78.5 x 72mm), 382cc (78 x 80mm), 387cc (78.5 x 80mm) and 402cc (80 x 80mm). There was also a short-stroke 249cc (74 x 58mm). It should be noted that all short-stroke engines, except the 249cc, had a larger bore in the mouth of the crankcase and a larger diameter sleeve and barrel, with the holding-down studs spaced further apart.

Richard Tree's 1967 350 Aermacchi, Snetterton, 22nd September, 1985. Note the special alloy tank, Yamaha TD/TR front brake, plus works flat-side fairing and front mudguard.

Mike Trimby, 402cc Aermacchi, Senior TT, early 1970s.

Finally, it was not until the birth of the classic movement that a full 500cc Aermacchi single was built by Syd Lawton, (not including the 1950s motocross engine with vertical cylinder), which is described in Chapter 13. Then, for the 1969 season Yamaha introduced its TD2 (250cc) and TR2 (350cc), which effectively made all the four-stroke opposition in these classes obsolete overnight, including the Ala d'Oro models. This set Aermacchi on the path to developing its own two-stroke twins, which are described in Chapter 11. However, for many, these machines did not replace Aermacchi's Golden Wing, the much-cherished over-head valve horizontal single, which by now had been around for the best part of a decade. Yes, the two-strokes were quicker, much quicker, but somehow they just didn't have the character to go with the speed. Also the two-strokes were far more difficult in set-up and reliability than the definitive versions of the four-stroke single. This had been developed to a stage where the word Aermacchi stood for robustness and strength. It was certainly a model which had gone from being one of the most fragile of racing machines when it had first appeared in the early 1960s. Credit must go to the factory development team and to Syd Lawton and Dick O'Brien for their independent work in the field.

Left; Stuart Morrell with his 350 Ala d'Oro at Snetterton Combine Reunion, Sunday, 19th September, 2004. Right; Snetterton, 26th May 2003, left to right, Andrew French (rider) and mechanics Bob Combley and his son Richard. The bike has a 1966 chassis with a later spec. engine, built and supplied by Dick Linton.

Syd Lawton

Syd Lawton was born on the 20th December 1913, in Tunstall, Stoke-on-Trent, but moved to Colwyn Bay, North Wales in 1925, where his parents were to run a boarding house. His father had been an engine-winder, operating the lift cage carrying miners to and from the coal-face. His mother was a school teacher.

At 17, Syd became a pharmaceutical apprentice and after five years, he joined the Prudential Assurance Company as a sales representative. However neither of these careers appealed so in 1937 he joined the Sheffield branch of the nationwide chain of motorcycle dealerships, Grays ,as a salesman/ mechanic. Within a few weeks he was promoted, becoming manager of his own branch in Manchester. In September 1938 Syd was moved to Southampton and stayed with Grays until war broke out, when with his mechanical knowledge, he was recruited into the aircraft industry as a fitter.

The end of the conflict did not see an immediate return to the motorcycle industry, but eventually Syd managed to pick up the threads of the game he loved. The Southampton branch of Grays had closed with the outbreak of war, so Syd became a salesman for a rival firm. At this time Syd was a Rudge enthusiast, and when he heard of a job lot of Rudge spares going for sale, he snapped them up. Initially, these were intended for his own use, but unwittingly the Rudge spares stock served as the nucleus of his own business, which developed over the next few months. Becoming 'restless' at his official workplace, in 1947, in partnership with a friend Pat Wilson, Syd took over a small garage in Freemantle, Southampton, where the two men concentrated upon repair work rather than sales.

The acquisition of the local Vincent agency in 1951, was, as Syd was later to recall, 'largely for Pat's benefit, he was a Vincent enthusiast, but I never really had interest in them.'

Syd had developed a great interest in racing. After witnessing his first Isle of Man TT in 1934 Syd caught the racing bug, and he bought one of Harold Daniell's Nortons in 1939. His first race was in a scramble, and he won on a 1928 Rudge with hand-gear change. He made his Island debut in September 1947, at the Manx Grand Prix, but retired with a split oil tank. Syd first rode in the TT in 1948, gaining a silver replica in the Junior race on an AJS 7R, loaned to him via Southampton dealers Jenkin and Purser. The other love of Syd's life was Beryl. They married in Shirley, Southampton on the 9th April 1941. The couple had six children, the oldest boy John, born in 1942. Then came Pamela, Barry, Sheila (sadly to die at only 34 years of age in a car crash in 1986), Christine and finally Ian in 1958.

Syd joined up with Tommy Woods at the beginning of 1952, when he had been suspended by the F1M for six months. Syd rode Tommy's bikes, including a very fast 250cc Moto Guzzi, Syd's performances that year were so good that he was signed by Joe Craig for the works Norton Team. However, practising for the 1953 TT, Syd crashed heavily at Creg ny Baa, breaking 37 bones, spending five months in Nobles Hospital, during which time gangrene brought him close to death at one stage.

John Hartle, with Syd, Spring 1967.

With his racing career at an end, Syd retired to Southampton determined to throw everything into the development of the business. 'Lets forget all the competition stuff, we'll concentrate on the bread-and-butter bikes from now on,' Syd told Pat Wilson, who agreed, and the business went from strength to strength.

Eighteen months later they purchased an old malt-house, for £6,000. This site remained as the Lawton and Wilson headquarters, even though expansion took place elsewhere.

After Vincent closed in 1955, Lawton and Wilson took on the Royal Enfield agency. Sporting involvement included the Rickman brothers riding Bullet 500 scramblers, and participation with Royal Enfield twins in the Thruxton 500 mile race from 1957 onward. Earlier (in 1955 and 1956), Syd had prepared Norton 88 Dominators for Thruxton. On the Enfields (from 1957 through to 1961) Bob McIntyre, Derek Powell and others led the race, only to strike trouble.

Then in 1962 Syd changed from Royal Enfield to Norton for the 500 miler and promptly won with the newly released 650SS model, ridden by Phil Read and Brian Setchell. The team then went on to a further two victories, from 1962 to 1964. Switching to Triumph for 1965 the success continued. Also in 1962 Syd purchased a new 250cc Aermacchi for son Barry to race. This was to lead to over thirty years involvement with the Italian marque. Subsequently, later in 1962 he became the Southern distributor. Then when Bill Webster died suddenly at Mallory Park in April 1963, Syd purchased the Aermacchi UK concession from Bill Webster's estate. On Monday, 27th May 1963 he was able to report, 'I have purchased the entire stock of spares and machines, both touring and racing. An agreement in principle for my firm to be the official concessionaires has been reached with the Italian Factory and only remains to be formally ratified. I shall be taking racing spares to the Island and will be staying at the Douglas Bay Hotel.'

Syd's input from this time is recorded elsewhere in this book, which of course includes the birth of the classic racing era from the late 1970s onwards. Syd was deeply involved with this movement and considered Richard Swallow as one of the very best riders he had ever had under his wing. Then in the mid 1990s Syd's health deteriorated and he died, aged 85, on the 17th January 1997.

Conflict of view

In the mid 1960s the first signs of conflict appeared between the Italian and the American factions within the organization about what development path should be taken. Ing Bianchi considered that to remain competitive at world level it would be necessary eventually to offer an overhead camshaft design in contrast to his earlier thoughts when designing the Chimera engine less than a decade before. So, with costs once again to take into consideration, Bianchi sketched out a double overhead cam cylinder head for the existing flat single. At around the same time, factory employee Celestino Piva designed and built his own dohc head. But the American management vetoed the plan and subsequently, even though two complete machines were built, the whole lot was sold, eventually becoming the property of Swiss engineer and enthusiast, Yves Liengme who had also been his country's Aermacchi importer for most of the 1960s.

The 'official' dohc head designed by Bianchi allowed the normal short-stroke 250 (72 x 66mm) engine to run up to 11,000rpm. The camshafts were driven by shaft and

bevel gears, as on a Manx Norton, with the shaft horizontal, in parallel with the cylinder. Hairpin valve springs were used. Compression ratio was a high 12:1 with coil ignition feeding a twin plug cylinder head and carburation taken care of by a 35mm Dell'Orto instrument. The original finish of the machine was silver-grey, or black and green and featured one-off tank and seat.

The privately-built Celestini Piva model was technically quite different from the official factory version, in having its camshaft driven by a series of three large helical gears in the cylinder head, located on the offside (right) of the engine.

The Bianchi and Piva layouts were also quite different in appearance. From a purely racing standpoint, it was a great pity that the double knocker didn't replace the pushrod cylinder head, but by that time, the American market was becoming the dominant factor, with almost 90% of production heading across the Atlantic. It must be understood that because of this, Harley-Davidson, rather than the European Aermacchi motorcycle, had to come first in Varese's planning, because the AMA ruling, which effectively barred double-overhead-camshaft engines from American racing, coupled with Harley-Davidson's insistence that the production roadster and racers use the same basic engine type, required pushrods, not overhead cams. This was the decisive factor.

Rickman and Drixl

John Hartle (Aermacchi-Metisse, 83) leads Dan Shorey (Norton, 76) through Holenstein-Ernstthal at the 1968 East German GP. Hartle was one of several top riders to use the Rickman frame, rather than the standard Aermacchi chassis.

A fine example of the Aermacchi-Metisse, a particularly smart racing motorcycle.

The Rickman-Metisse with fairing removed, a bigger machine than the standard Ala d'Oro and suited to taller riders.

The Metisse frame with its duplex front downtubes which passed over the top of the crankcases, rather than underneath as it had done on the Rickman design for the British-engined models. The idea for fitting the Aermacchi power unit had originally come from the Swiss rider/engineer, Othmar (Marly) Drixl. The first prototype appeared at the Earls Court Show in Autumn 1966.

Another important development was a new British Rickman Metisse frame kit for the 250/350 horizontal single-cylinder racing motor. The first prototype had been constructed for the Swiss rider, Othmar (Marly) Drixl, who had suggested the idea privately to the Hampshire brothers during the 1966 Earls Court Show. Drixl at that time worked at the Aermacchi factory, and initial testing of the new chassis was carried out by him at the Monza circuit.

Clearly, the new frame had been derived from the earlier Metisse, and was designed to accept British vertical engine units. The duplex down-tubes and cradle only needed relocating to house the horizontal Italian single-cylinder engine unit. All other Metisse components, tank, seat, wheels and front fork, were largely the same as on the other Rickman racers. Unlike the British-engined Metisse models, no separate oil tank was required for the wet-sump Aermacchi.

The uniformly diametered tube Metisse frame was certainly larger and more robust looking than the

Left, most Aermacchi-Metisse models were supplied with Rickman-Lockheed front brake. As shown on the right, the exhaust was well tucked in.

Guy Martin with Rickman-Metisse and fairing. The fairing on the right is for a Ducati single.

single tube, spine-backbone Aermacchi original, and was constructed from Reynolds 531 tubing. Other details in its specification included Timken roller bearings at the steering head, an eccentric swinging arm spindle to facilitate chain adjustment, and bright nickel-plate finish. The front fork rake angle was increased from 27 to 29 degrees. There was a longer swinging arm and wheelbase, and Rickman claimed there was increased torsional stiffness due to the double cradle layout and a lower (10mm) engine position.

A year on, in October 1967 Drixl was working in close co-operation with Italian engineer Egidio Baroni; the partnership of Drixl with the Milan-based Baroni created what was in many ways a scaled- down Metisse chassis. Drixl considered the Metisse frame to be 'too big'. Therefore, he had designed a lighter and lower version, incorporating a number of modifications.

The prototype was built in Baroni's Milan factory and brought to Britain where it was tried at Snetterton by a number of top riders, including Dave Degens and Ray Pickrell.

Although it clearly followed Metisse lines, the Drixl-Baroni frame (marketed as the Drixton) was in fact much lower than the Rickman. For chain adjustment,

Brian Lee geting down to it, on his 382cc Aermacchi-Metisse.

fore-and-aft location of its swinging arm spindle could be varied by eccentric plates. Bolted to the frame, the footrests were easy to replace if damaged in a crash.
The Drixton's short front fork looked like a Ceriani, but was actually manufactured by Baroni. Well-known at the time in Italian motorcycling circles, Baroni had previously built the forks used on several racing bikes, including the FB Mondials ridden by Taquinio Provini in 1957, and the Morini on which Giacomo Agostini made his international debut in 1963.
In the production model, further weight-saving was achieved. The rear sub-frame was lighter and lower, and the swinging arm was a reinforced rectangular section with chain adjustment by light-alloy 'snail' plates on the wheel spindle, instead of by eccentrics on the swinging arm spindle. In the first year, twenty Drixton-Aermacchis were built and sold.
To accommodate the demand for engines, the Aermacchi company reached an agreement with people such as Syd Lawton and Othmar Drixl to supply new engine assemblies. From the end of 1967 it was also possible to purchase them through local importers. In Britain in 1968, a 350 unit cost £265, and a 250 cost £242.

Steve Jolly at Ballaugh Bridge, 1968 Junior TT aviating his Aermacchi-Metisse.

Francis Beart

Veteran tuner Francis Beart (right) with rider Jack Findlay, Francis continued with the Italian marque until his retirement in the early 1980s

The name F. L. Beart first found its way into the motorcycling press during the 1930s, when the Brooklands, Surrey circuit was the mecca for speed-men the world over. Francis himself was a regular competitor for a number of years. In 1937 he moved into a small workshop, described by 'The MotorCycle' as a 'shack' on the circuit and set about making other people's engines go more quickly, instead of riding his own. Building up a business from scratch calls for dedication and hard work. But Beart went even further, for he almost lived at his tuning establishment and worked hours longer than any man should be expected to. However, in terms of results if not financial return, he could claim complete success. Before Brooklands' gates were finally closed in 1939 his efforts had paid off in a big way, twelve track and three world records had been claimed by Beart-prepared machinery.

Then came the war and for six long years there was no motorised sport, but from mid-1945 peace-time pursuits, including racing, began to return. Francis Beart was soon to find himself even busier, tuning for many of the greats of the road racing game, including Johnny Lockett, Ken Bills, Cromie McCandless, Manliff Barrington, Denis Parkinson, Dan Crossly, Peter Romaine and many, many more. He became particularly involved with the Norton marque, and in fact acted as the company's official representative at Daytona, where Nortons tuned by him swept the board in the American classic for three years, 1949, 1950 and 1951.

Beart temporarily quit motorcycles for cars. The early 1950s were a boom era for light-weight 500cc Formula 3 cars. Demand for his services from the four-wheel brigade was overwhelming. But in 1958, with Formula 3 in decline, he rekindled his interest in motorcycles and returned to the fold.

This came via an approach to 'breathe' on the pair of Nortons to be raced by Ernie Washer in the Manx Grand Prix. This combination of Beart's tuning ability and Washer's track-craft resulted in victory in the 500cc Senior race at record speed, and a new lap record was set. Washer might well have won the Junior too, had he not missed a gear and bent a valve. Then came a whole host of Manx GP successes, including Peter Middleton's victory in the 1959 Junior and Ellis Boyce's victory in the same event the following year. Other riders who benefitted from Beart's tuning magic included Mike Hailwood, Joe Dunphy, Terry Shepherd and Bob Anderson.

Assisted by Phil Kettle, the Guildford-based tuner continued to prepare Nortons until well into the 1960s. Eventually Francis Beart turned his attention full-time to Aermacchi machinery, after a brief

flirtation with Greeves two-strokes. Like all Beart machines, these bikes were finished in a distinctive light green livery.

The first Beart-entered Aermacchi success came in the 1969 Junior TT in the Isle of Man, when the Australian Jack Findlay rode a 350cc Ala d'Oro into third spot. The machine bore all the hallmarks of Francis Beart, with its detailed preparation and weight-saving features. Just some of the fine work included the myriad of drilled and lightened components (such as engine mounting plates, rear sprockets, mudguard stays, brake torque arms etc), whilst major weight-pruning came via the fitting of the smaller Oldani brake and front forks normally associated with a 250.

Besides Findlay's TT success, the same machine with Clive Brown aboard won the 1970 Junior Manx Grand Prix. Francis Beart continued his involvement with Aermacchi until his retirement in the early 1980s.

The Beart Aermacchi without streamlining to reveal amazing attention to detai.

Ala d'Oro - year by year changes

1961 246cc long stroke racer, 66mm bore x 72mm stroke, wet-clutch 4-speed gearbox, 30mm carb, cast piston, long conrod, DS2 cam, 10mm diameter, stellite tappets, 9,000rpm.

1962 246cc long stroke racer, 66mm bore x 72mm stroke, wet-clutch 4-speed gearbox, 30mm carb, cast piston, long conrod, DS5 Cam, 10mm diameter stellite tappets, 9,200rpm.

1963 246cc long stroke racer, 66mm bore x 72mm stroke, wet-clutch 5-speed gearbox, 30mm carb, forged piston, short conrod, 'A' cam, 10mm diameter stellite tappets, 9,400rpm.

1964/5 248cc short-stroke racer, 72mm bore x 61mm stroke, wet-clutch 5-speed gearbox, 30mm carb, forged piston, short conrod, needle rollers to inside of cam, 'A', 'AF' or 'F' cam, 10mm diameter stellite tappets, 8,000rpm.

1964/5 344cc racer, 74mm x 80mm stroke, wet-clutch 5-speed gearbox (a few have 4-speed close ratio gears), 30, 32, 35mm carb, forged piston, long conrod, 'A', 'AF' or 'F' cam, 10mm diameter stellite tappets, 8,000rpm.

1966 248cc racer, 72mm bore x 61mm stroke, dry-clutch, 5-speed gearbox, wider gears, large diameter flywheels but no outside flywheel, 30mm or 35mm carb, forged piston, short conrod, needle rollers to inside of cam, 'A', 'AF' or 'F' cam, small diameter cam where it fits through camshaft outrigger bearing plate, small diameter camshaft/ points oil seal, small diameter LHS

engine mainshaft oil seal, RHS engine mainshaft is located in the bearing by a distance piece and circlip on inside face (nearest to the flywheel), RHS engine mainshaft is located in through the camshaft outrigger plate as per all previous models. All clutch and gearbox parts from 1968 onwards are the same (apart from the availability of different internal ratios), 9,800rpm.

1966 344cc racer, 74mm bore x 80mm stroke, spec as per 1966 250 Racer except long or short conrod, 35 or 38mm carb, larger diameter flywheels than later 250/350, longer holding-down studs and pushrods than 250, 8,000rpm.

1967/72 248cc racer, 72mm bore x 61mm stroke, dry-clutch 5-speed gearbox, small diameter outside flywheel, internal flywheels are not full, but bob weights, 30 or 35mm carb, forged piston, short conrod, 'H' cam as standard but where the engine has been improved 'L', 'LI' or DS 175 cams may be fitted, 10mm diameter chrome tappets as standard but with later cams 12mm diameter chrome tappets give greater reliability. From 1967 onwards 250 and 350s have different timing cover with two external oil pipes, the second pipe going to the bottom of the timing cover with the oil fed up through to a central bronze bush for the big-end feed. A large diameter camshaft/points oil seal is fitted. Also in 1967 the crankcases were modified from new with two ball main bearings on LHS crankshaft, the additional bearing being fitted into a bolt on main bearing housing located by four long allen screws. The LHS crankshaft is longer to accommodate the extra main bearing, and the outside flywheel is fitted on the end of this shaft, located by a large size key and left-hand threaded nut, the only one on the bicycle. The LHS oil seal is larger diameter than previous model's and fits over the outside flywheel boss. The seal is the same as the dry-

Opposite, superbly maintained 1966 350 Ala d'Oro seen at the Bob McIntyre Memorial Meeting, East Fortune, 19th - 20th June 2004. This page, Philip Sharpe's very nice 350 Ala d'Oro, Scarborough International Gold Cup Road Races, 22nd - 23rd September 2001.

clutch gearbox bearing seal, and the 35mm Ceriani fork oil seal. The flywheels are no longer located in the crankcase by use of the circlip on RHS mainshaft, but from 1967 on, simply by use of shims on the engine mainshaft usually on LHS, and never by use of shims less than 0.01 inch as they would break up. Correct end float 0.008 inch to 0.01 inch.

From 1967 onwards the standard 5-speed dry-clutch gears are 'B' ratios, though a few engines are fitted with 'A' ratios, which are closer.

Most special 250s and genuine works engines have special eccentric rockers; this also applies to the short-stroke 350, 382 and 402 works engines. The special 250s usually have 35mm carb in place of the standard 30mm, though on at least one ex-works machine, a 38mm carb is fitted,10,200rpm.

1967/72 344cc racer. The specification for all production 350s from 1967 on is as for the 250, except for carburettor size, which may be 35, 38 or 40mm. Also there are short and long conrod motors. Prior to and including 1967, most longrod 350s have a packing piece under the barrel approx 45 thou thick. From 1968 on the longrod 350 had this extra thickness added to the base of the barrel. Original 1967 350s had 'F' cams and 10mm diameter stellite tappets. In 1968 the 'N' cam with 10mm diameter chrome tappets was introduced and subsequently 'N5' and 'N6' cams. Generally, where engines have been updated by the fitting of later cams, 12mm chrome tappets have also been fitted to give greater reliability. Note that when fitting later cams it is also necessary to fit type 23 rockers which have a lower lift ratio.

1968/72 349cc racer. short-stroke works special, 77mm bore x 75mm stroke, 38mm carb, forged piston,

Aermacchi by Mick Walker

N6 cam, 12mm tappets, compression ratio 11.5:1, max power 8,400 to 8,600rpm, 42bhp, ignition 30°, otherwise as per 344cc model.

1968/72 348cc racer, 38 or 40mm carb, forged piston, DSS L/1 cam, 12mm diameter tappets, compression ratio 11.5:1, max power 8,800 to 9,000rpm, 42bhp, ignition timing 36°, primary drive ratio 2.09:1, long conrod, otherwise as 344cc model. Ultra-short-stroke works special, 78.5mm bore x 72mm.

1968/72 382cc racer, works special, 78mm bore x 80mm stroke, special 20mm diameter gudgeon pin and conrod to suit, 38 or 40mm carb, forged piston, L11 or- special N6 cam, 12mm diameter tappets, compression ratio 11.25: 1, max power 8,000rpm, 42bhp, ignition timing 35° 36°, otherwise as per 344cc model.

1968/72 387cc racer, 78.5mm bore (as per ultra short-stroke 348cc) x 80mm stroke, 38 or 40mm carb, forged piston, N6 cam, 12mm diameter tappets, compression ratio 11.25:1, max power 8,000rpm, 42bhp, ignition timing 35°, otherwise as per 344cc model.1968/72 402cc racer, works special, 80mm bore x 80mm stroke, 38, 40 or 42mm carb, forged piston, N6 cam, 12mm diameter tappets, compression ratio 11.25:1, max power 8,000rpm, 42 to 44bhp, ignition timing 33°, otherwise as 344cc model.

1968/72 249cc racer, works special, 74mm bore x 58mm stroke, 38 or 40mm carb, forged piston as late 344cc, DS 175 cam, max power 11,000rpm.

NOTE All short-stroke engines, except the 249cc, have a larger bore in the mouth of the crankcase and larger diameter sleeve and barrel, with the holding-down studs spaced further apart.

Opposite, the Esthwaite Aermacchi-Metisse campaigned so successfully by Tony Myers at CRMC and IHRO events in recent years.

This page, Drixon 350 Aermacchi, CRMC Cadwell Park, 5th October 2002.

9 European Roadsters

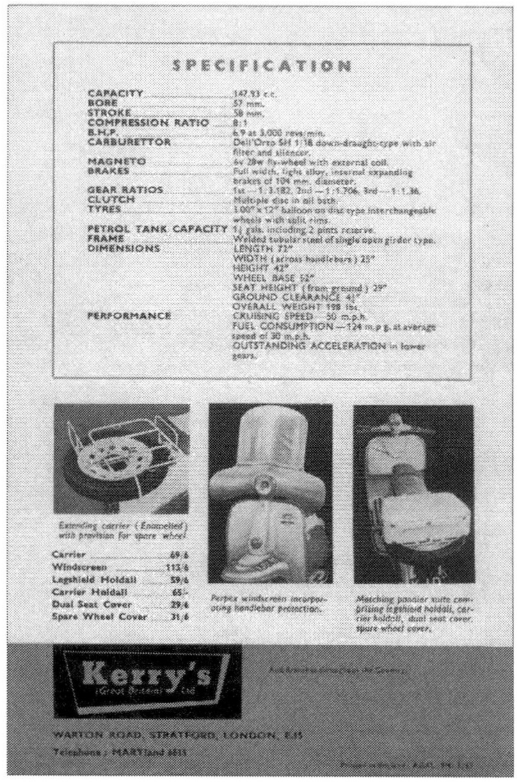

In the early 1960s Aermacchi built the Brezza scooter, powered by a 148cc (57 x 58mm) fan-cooled 2-stroke engine with 3-speeds, flywheel magneto, 12 inch wheels and welded tubular steel frame. It was imported into the UK in 1962 and 1963 by London based Kerry's.

For much of the period after Harley-Davidson gained a 50% share in the Varese Aermacchi motorcycle operation, the majority of production crossed the Atlantic for sale in the States (see Chapter 6). However, there were still European sales, and these bikes were often quite separate models, rather than simply American-market models with a fresh coat of paint and a different handle-bar bend.

In fact there were eight European-market models listed in the 1960 range. These were road-going machines, as opposed to competition models. They were the Chimera in 175 and 250cc engine sizes; the sporting Ala Rossa (175) and Ala Verde (250); the Ala Bianca (White Wing) 175 and Ala Azzura (sky blue wing) 250 touring machines; plus the Zeffiro 125 and 150 two-stroke commuters which were essentially large-wheeled scooters. The top performer was the Ala Verde, which the factory claimed was capable of 86mph with full road equipment and a standard Silentium silencer.

1963 Brezza Scooter

Engine:	Fan-cooled piston port horizontal single cylinder two-stroke, alloy head, cast-iron barrel unit construction
Bore:	57mm
Stroke:	58mm
Displacement:	147.93 cc
Comp ratio:	8:1
Lubrication:	Petroil mix
Ignition:	6 volt, flywheel magneto 28 watt
Carburettor:	Dell 'Orto SH 1/18
Primary Drive:	Helical gears
Final Drive:	Chain
Gearbox:	3-speed, hand-change (twistgrip)
Frame:	Pressed-steel construction
Front suspension:	Leading axle telescopic fork
Rear suspension:	Swinging arm, twin shock absorbers
Front Brake:	Full width, light alloy, 104mm
Rear Brake:	Full width, light alloy, 104mm
Tyres:	3.00 x 12 front and rear

General Specifications

Wheelbase:	52 inch (1320mm)
Ground clearance:	5 inch (127mm)
Seat height:	29 inch (736mm)
Fuel capacity:	1.5 gallon (7 ltr)
Dry weight:	198lb (90kg)
Maximum power:	6.9bhp@5000rpm
Top speed:	50mph (80kph)

The following year, 1961, saw the production figures at Varese show a very healthy increase over the previous year's, helped in no small part by the large number of machines exported to the United States to be sold through the comprehensive Harley-Davidson dealer network.

As mentioned elsewhere, dealer and former racer Bill Webster had signed an agreement with Aermacchi Harley-Davidson, to sell selected models in Great Britain, namely the Ala d'Oro 250 racer, Chimera 250, Ala Verde 250 and the Zeffiro 150. Webster had formed a new company, Italian Imports Ltd, based in Snow Hill, Nantwich, Cheshire.

For 1962, Webster brought in five models; the Ala Rossa, Ala Verde, Chimera 250, the Ala d'Oro racer, and the Wisconsin which was essentially the American version of the Ala Azzura 250 touring model.

Besides these, the factory also produced the Ala Bianca, 175 tourer and the new

AERMACCHI HARLEY-DAVIDSON
I motocicli perfetti per ogni uso e per ogni borsa
Les motocyclettes parfaites pour tout usage et possibilité
The perfect set of wheels for every purse and every purpose
Zeffiretto
Zeffiretto Sport
Ala Rossa G.T.
Ala Rossa S.
Brezza
Ala Azzurra
Ala Verde
Wisconsin
Ala d'Oro

1960 175 Ala Bianca 250 Ala Azzurra

250 in brackets where different

Engine:	Air-cooled ohv single with horizontal cylinder; alloy head and cast-iron barrel; built-up crankshaft with roller bearing big-end; vertically-split aluminium crankcases; unit construction
Bore:	60mm (66mm)
Stroke:	61mm (72mm)
Displacement:	172.3 cc (246.2 cc)
Comp ratio:	7:1
Lubrication:	Gear-type pump; wet sump
Ignition:	6 volt battery/coil, crankshaft-mounted generator
Carburettor:	Dell 'Orto
Primary Drive:	Helical gears
Final Drive:	Chain
Gearbox:	4-speed, foot-change
Frame:	Spine type
Front suspension:	Telescopic fork
Rear suspension:	Swinging arm; twin shock absorbers
Front Brake:	Full width alloy drum, SLS
Rear Brake:	Full width alloy drum, SLS
Tyres:	Front 2.50 x 17, Rear 2.75 x 17 (3.00 x 17)

General Specifications

Wheelbase:	52 inch, 1320mm
Ground clearance:	5 inch, 127mm
Seat height:	30 inch, 760mm
Fuel capacity:	3.75 gallon, 17 ltr (18 ltr)
Dry weight:	242lb,110kg (245lb.111kg)
Maximum power:	9.25bhp@65000rpm (13.7bhp@65000rpm)
Top speed:	64mph, 105kph (78mph, 125kph)

for 1962 Brezza scooter, which was intended to replace the Zeffiro 125/150 models, which had finally been axed after many years of service.

The Brezza was a conventional small-wheeled scooter in the style of Lambretta and Vespa. This meant a typical scooter of the period with 12 inch pressed steel wheels, alloy brake hubs, a dual seat, and single vertical cylinder 2-stroke engine with piston-porting and fan-cooling. It had a capacity of 147.93cc (57 x 58mm) and a 3-speed gearbox.

This was also imported into Britain, not by Webster's Italian Imports, but by one of the original motorcycle industry names, Kerry. This famous old pioneering name had been reformed to sell the Kerry Capitano moped which was technically interesting for its use of a nylon moulding for its peak-type front mudguard. This firm traded as Kerry's Limited and was based in Warton Road, Stratford, El5. Unfortunately, by the time the Brezza was launched, scooter sales across Europe were in decline so few Brezzas were sold in the UK, whereas a considerable number of Capitano mopeds were purchased by British buyers during the early 1960s. The price in November 1962 was £139 19s 6d, including UK taxes.

New for 1963 was a 49.61cc (38 x 42mm) two-stroke engine with 3-speed gearbox. This was used to power the Zeffiretto, a 'step-thru' moped as well as the Zeffiretto Sport ultra-lightweight motorcycle. Both were sold in Europe and North America.

Actually, although most enthusiasts would associate the well-known ohv horizontal single cylinder family as Aermacchi's mainstay during the 1960s, the two-stroke also played an increasingly important role as the decade unfolded. Central to this was an all-new 125-class two-stroke. As before, it featured piston-port induction and a single vertical cylinder. Compared with the 50cc range, the newcomer was to prove an altogether more successful machine which was to lead to a whole family of new models.

Swiss Hartman Special, using a 250cc Aermacchi engine unit with five-spoke wheels, Ceriani forks and steeply inclined rear shock absorbers.

This machine, designated the Aletta, featured a 123.15cc (56 x 51mm) engine designed by the German engineer, Peter Durr. It was also to be known as the M125 Rapido in North America. The engine, as was usual with Aermacchi, was of full unit construction design, with integral gearbox (housing four ratios) and helical gear primary drive. Running on a compression ratio of 7.3:1, the ML put out 10bhp and could reach 105 kilometres per hour (65mph). Other details included a wet multi–plate clutch, enclosed telescopic front forks and twin rear shock absorbers. Ignition was by Dansi flywheel magneto, whilst there was a spine frame, the alloy cylinder head being bolted to the frame at the rear. There was also a large circular metal container at the rear of the carburettor, which housed the air filter. Another feature was the use of polished aluminium full-width brake hubs, 135mm front and rear; the rear contained a rubber cush drive vane. All-in-all, the new 125 two-stroke was a welcome addition, offering as it did, a bridge in the Aermacchi model range between the 50cc and the 250cc models as by now the 175 series had been discontinued.

A street scrambler version arrived for the 1968 season (sold as the ML 125 in the States). Another new two-stroke engine appeared for 1968 in the shape of the M65. This was built initially as a conventional motorcycle both as the M65 and

The 1967 Rapido (also known as the ML) was an entirely new ultra-lightweight motorcycle powered by a 123.5cc (56 x 50mm) 2-stroke single cylinder engine.

the M65S Leggero. Later, but only in North America, it was built in mini-bike form, inspired by Honda's successful 'Monkey Bike'.

Meanwhile from 1967 the 250 ohv horizontal single cylinder range had been updated; not only did it now sport a 5-speed gearbox, but new shorter stroke engine dimensions of 72 x 61mm, giving a new displacement of 246.4cc. In the American market all the 250cc horizontal singles were referred to under the heading of Sprint, but in Europe the various versions had individual names, as already outlined. However, from the introduction of the 5-speed 250, only two models were offered, the Ala Verde 5 Marce (speed) and the Ala Blu (blue wing) GT 5 Marce. Actually, the two machines used an identical engine specification, with a compression ratio of 8.5:1 and 24mm Dell 'Orto UBFBS carburettor.

Compared to the earlier Ala Verde the new version was much the same, but with a side stand in place of the centre stand, polished, bare aluminium fork sliders (bottom legs), rubber fork gaiters, a new, larger dual seat, exposed spring chrome-plated shock absorbers, larger section and 18 inch instead of 17 inch rims. Also the tank, frame, side panels, fork yokes, shock absorber bodies and

1969 Aletta (Rapido USA)

Engine:	Air-cooled piston-port single cylinder two-stroke alloy cylinder head cast-iron cylinder barrel, unit construction, vertically-split aluminium crankcases
Bore:	56mm
Stroke:	50mm
Displacement:	123.15 cc
Comp ratio:	7.3:1
Lubrication:	Petroil mix
Ignition:	Dansi flywheel magneto, 6 volt
Carburettor:	Dell 'Orto
Primary Drive:	Gears
Final Drive:	Chain
Gearbox:	4-speed
Frame:	Spine type
Front suspension:	Telescopic forks
Rear suspension:	Swinging arm, twin shock absorbers
Front Brake:	5LS drum, 135mm
Rear Brake:	5LS drum, 135mm
Tyres:	2.50 x 17 front and rear

General Specifications

Wheelbase:	48.9 inch, 1242mm
Ground clearance:	6.7 inch, 17mm
Seat height:	29.5 inch, 749mm
Fuel capacity:	2.2 gallon, 10ltr
Dry weight:	195lb, 88.5kg
Maximum power:	10bhp
Top speed:	66mph, 105kph

The Aletta 125, produced 10bhp and 66mph, it was developed from the Rapido, and debuted in 1969.

Aermacchi by Mick Walker

A restored 350 GTS out on test with Motorcycle Enthusiast magazine in 1985.

Instrument console with matching Veglia clocks, reset trip and warning lights, 350 GTS

Twin leading shoe front brake with 180mm full-width hub. Forks equipped with fork gaitors, abbreviated mudguard and chrome steel rim.

headlamp brackets had a colour change, silver was used for the headlamp shell, mudguards, rear light support and a third dividing line on the side of the tank. Factory sources claimed a top speed of 140kph (88mph), from a power output of 18bhp at 6750rpm, and a dry weight of 120kg.

The Ala Blu GT 250 was finished in silver (tank, mudguards and headlamp shell) and black (frame, panels, chain-guard, upper fork components, including headlamp brackets). Except for an increased dry weight of 129kg and a lower maximum speed of 135kph (84mph), the general specification was unchanged. However, the styling was considerably different, with higher and wider handlebars, a new saddle and a peanut-shaped fuel tank, plus a polished aluminium binnacle, which held both the Veglia speedometer and rev-counter. Strangely there was no rev-counter as standard fit on the more sporting Ala Verde.

In 1969 came the final development in the Aermacchi ohv horizontal single, a new 350cc model; even though such an engine size had been available for the Ala d'Oro road racer for some time (see Chapter 8).

1970 350 GTS (Sprint USA)

Engine:	Air-cooled ohv single with horizontal cylinder alloy head and barrel, built-up crankshaft with roller bearing big-end, vertically split aluminium crankcases; unit construction
Bore:	74mm
Stroke:	80mm
Displacement:	344 cc
Comp ratio:	9:1
Lubrication:	Gear-type pump; wet sump
Ignition:	6 volt 12 ah battery/coil; 60 watt generator
Carburettor:	Dell 'Orto VHB 27A
Primary Drive:	Helical gears
Final Drive:	Chain
Gearbox:	4-speed; foot-change
Frame:	Spine type
Front suspension:	Ceriani telescopic fork
Rear suspension:	Swinging arm, twin shock absorbers
Front Brake:	Drum 2LS, 180mm full-width alloy hub
Rear Brake:	Drum SLS, 180mm full-width alloy hub
Tyres: Front	3.00 x 19; Rear 3.50 x 18

General Specifications

Wheelbase:	53 inch, 1350mm
Ground clearance:	5.5 inch, 139mm
Seat height:	30.4 inch, 774mm
Fuel capacity:	2.6 gallon, 12ltr
Dry weight:	308lb, 140kg
Maximum power:	25bhp@7,000rpm
Top speed:	83mph, 150kph

European market version of the Sprint, the GTS, 344cc engine, branched exhaust and twin leading shoe front brake.

The GTS was very much for touring rather than for sport. The widely spaced four speeds dulled performance potential.

With bore and stroke dimensions of 74mm x 80mm, the engine size was 344cc (the same as the customer racer). The factory quoted power output as 25bhp at 7,000rpm, whilst the dry weight was 311lb. Other details included a compression ratio of 9:1, a 4-speed gearbox, dry-clutch (again like the racer) and internal gear ratios of; 1st 2.92:1, 2nd 1.76:1 and 3rd 1:1. The carburettor was a 27mm square slide Dell 'Orto VHB. But most interestingly, not only did the newcomer, marketed in Europe as the GTS 350, have a branched exhaust header pipe (from single exhaust port) with two silencers, but also a new, one piece rocker cover, which encased all the valve gear, giving the top end of the engine a much chunkier appearance (many were delivered with the conventional rocker box arrangement). The front brake, still a full- width aluminium drum, now boasted a twin leading shoe operation. Like the Ala Blu GT 250, the 350 GTS had a rev counter as standard equipment, this being housed in the same aluminium binnacle above the headlamp which also mounted the speedometer as a matching pair.
Official factory sources quoted a maximum speed of 155kph (97mph). In fact, the 'new' GTS was virtually identical to the American SS Sprint model released the same year, it even had 'Sprint' side panel badges! There is no doubt that the twin Lafranconi silencer restricted outright performance, but it made the GTS quieter

1970 250 Ala Verde sport.

1972 350 TV

Engine:	Air-cooled ohv single with horizontal cylinder, alloy head and alloy barrel with iron liner, built-up crankshaft with roller bearing big-end; vertically-split aluminium crankcases; unit construction
Bore:	74mm
Stroke:	80mm
Displacement:	344 cc
Comp ratio:	9:1
Lubrication:	Gear-type pump; wet sump
Ignition:	6 volt battery/coil
Carburettor:	Dell 'Orto VHB 30AS
Primary Drive:	Gears
Final Drive:	Chain
Gearbox:	5-speeds, foot-change
Frame:	Spine type
Front suspension:	Ceriani telescopic fork
Rear suspension:	Swinging arm, twin shock absorbers
Front Brake:	Drum 2LS
Rear Brake:	Drum SLS
Tyres:	Front 3.00 x 19; Rear 3.50 x 18

General Specifications

Wheelbase:	53 inch, 1,350mm
Ground clearance:	5.8 inch, 147mm
Seat height:	30 inch, 762mm
Fuel capacity:	2.8 gal, 12.75ltr
Dry weight:	298lb, 135kg
Maximum power:	30bhp@7,200rpm
Top speed:	97mph, 159kph

Neat lines of the 1971 350TV - almost an Ala Verde for a new decade.

350TV handbook. The TV had five-speeds, the GTS and later N only four ratios.

than it would otherwise have been. Also unlike the latest 250s, the crankcase outer engine casings were highly polished, rather than left matt, as the rest of the engine unit. There was also a circular detachable cover giving immediate access to the clutch. A 'Europe only' model of the horizontal ohv single was the 350 TV sportster. Launched for the 1971 model year, this employed the 344cc dry-clutch engine, but now with a 5-speed gearbox. A larger 30mm Dell 'Orto VHB carburettor with

1972 350TV with blue and silver paint job.

One of the final 350N (Sprint) models sold in Italy during 1972 by Egatti of Brescia . Now owned by Peter Fleming. With 4-speed gearbox and rubber-mounted handlebars -the bike still has its original suppliers stickers

open bell-mouth (the GTS/Sprint having a comprehensive air filter) was fitted. But there were still 6–volt electrics, with a 60-volt alternator. Again, the branched exhaust header pipe and twin silencer remained, as did the 9:1 compression ratio. The factory claimed 29bhp at 7,500rpm and a top speed of 100mph (160kph). The 350 TV employed a 14-tooth gearbox sprocket and 32-tooth rear-wheel sprocket, plus flat sports handlebars, separate Veglia speedometer and rev counter with no binnacle of the type found on the GTS. It had alloy wheel rims, exposed stanchion front forks (rubber gaiters on the GTS), a shorter hump-back seat, larger fuel tank with cut-outs for the rider's knees, and exposed spring, rear shock absorbers.

TV buyers had the choice of orange, green or blue paintwork for the tank, side panels and mudguards; the orange in particular adding to the sporting image, but there was no red and white as had been the traditional colour scheme for both the Ala d'Oro racer and Ala Verde/ Rossa sportsters for so long. The TV was sold from the end of 1970 until mid-1972.

One of the 350 GTS (Sprint) machines was imported into Britain in 1971 by London dealer Fred (Oily) Wells, for evaluation. The bike was subsequently tested by Dave Minton for 'Motorcyclist Illustrated' magazine. He reported, 'Starting was difficult and often painful due to the impossible left side crank that displayed a strong tendency to hinge from under the foot on maximum depression. And the front mudguard was cosmetically inefficient. It was almost as though the manufacturers, dedicated racing men, regarded their roadster bread-and-butter models as a necessary evil, and distastefully slow stuff. Racer you say?, That's no racer. It's got lights and stands and a dual seat, and high bars and this and that. So come off it. Anyway, it's not that fast. I tell you, it is more than any other roadster that has ever come the way of 'Motorcyclist Illustrated'. This Aermacchi... (note no mention of Harley-Davidson!)... is a racing motorcycle accepting its hideous roadster lot with the grace of an aristocrat.'

The main heading of the Minton Test summed things up best, 'Irrepressible Racer-Not all the publicity men, stylists, market researchers or unsympathetic public in the world could ever stifle the racing genetics of this Aermacchi.'

The final days of the Varese factories models from the end 1972 onwards are covered in Chapter 12, as these were very much Harley-Davidson days rather than Aermacchi.

Peter Fleming's Macchi collection, left, the 1969 344cc Ala d'Oro he has owned for the last 30 years, purchased in May 1975 (raced by Graham Sharpe) and his 1972 350N 4-speed roadster.

The Dutch Market

The Netherlands first contact with Aermacchi dates back to 21st September 1950 and had nothing to do with motorcycles. It concerned a contract between Fokker and the aircraft manufacturer Macchi to build under licence the Fokker S-11 trainer. The Italian type number for this machine was M416. 120 machines were built in Varese. Aermacchi motorcycles were not imported into the Netherlands until 1957. The Dutch trading company Prosperita announced, the Chimera 175cc, 'The Italian wonder motorcycle has just arrived in the Netherlands. The ultimate motorcycle!' This model did no better in the Netherlands than in Italy. The next importer, Nimag, in Leidschendarn was more successful. They offered the complete range of models, the most popular being the Ala Verde. The price in 1961 was fl.2495 (£250). The Ala d'Oro was also offered in both street and sport versions.

In spite of the sporting exterior and good reception in the press, the normal street models such as the Wisconsin, Ala Rossa, Ala Azzura, Ala Verde and Chimera were not sales hits in the Netherlands. The quantities sold were so small that they can't be traced any more, as the distributors of the time lumped together the sales figures of all types of machine sold. The Brezza scooter was imported but only as a show model. The mopeds Zeffiretto and Zeffiretto Sport were not imported. The Ala d'Oro model had the most success as this was an affordable lightweight racer with good performance and roadholding.

Several Dutch racing names can be linked with Aermacchi over the years, including Jan Huberts, Han Leenheer, Joop Vogelzang, Theo Louwes, Leo Bovee and André de la Porte. Aermacchi first appeared on the Dutch racing scene in 1960 when an Ala d'Oro 250cc was ridden by J van Leusden. In 1963 there was an increase in Ala d'Oro competitors, including H. IJpma, W. Buitenbos, H. Veenstra, F. Holtkamp and Andre de la Porte who was seen for the first time on the starting grid. De la Porte in particular, a racing joumalist who worked for the weekly Motor, did much to promote the name of Aermacchi. This was so much appreciated in Varese that he was offered an experimental double-knocker engine with the agreement that he would become a semi-works rider. This did not happen as after the purchase of the engine, all Works interest in development was stopped. There was quite some secrecy around these dohc Oros, although the names of Tonti and Piva were mentioned in private circles. In the USA, Harley experimented with two such engines in dirt track frames, according to Dick O'Brien, the legendary former Milwaukee factory race boss.

In 1963 Rob Fitton and Ivan Sauter participated in the International races in Tubbergen, as well as Han Leenheer and Joop Vogelzang. H. IJpma became junior champion in 1963. In the six hour race at Zandvoort an Aermacchi Ala Verde ridden by Andre de la Porte won the 250cc class. In 1963 the price of an Ala Verde had risen to fl.2695 (£287) and if you wanted a five-speed gear box instead of a four, the price was fl.2870 (£294). The Oro then cost fl4195 (£420).

In 1964 Han Leenheer became the new senior champion in the 250cc class. New names in that year were H. Alkema, P. Vis and Th. Lablans.

1965 was a successful year for P. van de Goorberg, who won the Dutch championship at Etten, with F. Holtkamp winning at Tubbergen. Also that year the 350cc Ala d'Oro was offered by Nimag, 'top speed over 200 km per hour'. Then, during the international road races in 1966 at Tubbergen, Giuseppe Vizenzi gained a third place, with Han Leenheer placed 8th. At the national races at Zandvoort Jan Kostwinder was 3rd on an Aermacchi. The 250cc class of the international races in Tubbergen in 1967 was won by the Austrian Rudi Thalhammer and the 350cc by Karl Hopper, a double for Aermacchi. In

the national events few successes were booked.
At this point, the two-stroke machines made an appearance, and a new name appeared - Jaap Mooien. In 1968 he was 4th in the international races at Thburg. Also in that year Leo Bovee appeared for the first time on an Aermacchi as well as Theo Louwes and Henk de Wever. The 350cc race in Tubbergen was won by Karl Hopper on an Aermacchi.
In 1969 Jan Bestebreurte appeared on an Oro and later was to ride a Linto. During the international Olofraces Jerry Lancaster was third on a Drixton Aermacchi in the 350cc class, with Theo Louwes eighth on a similar machine. At this same event Louwes was 5th in the 250cc class. At Tubbergen, Lewis Young won the 350cc class, and in the national race Leo Bovee made 6th in the 250cc and 4th in the 350cc. During the international races in the same year at Oldebroek a two-stroke Aermacchi 125cc was seen for the first time, gaining a fourth place for Pierre Aubry. At Hengelo, Leo Bouvee took third place in the national 350cc class.
In 1970 at the international Olofraces Jim Curry came a 3rd place in the 350cc class on the fastest Aermacchi, Theo Louwes was 7th and Leo Bovee 15th, with Gyula Marsovzky coming in 5th on a Linto in the 500cc class. At Uden a new Aermacchi rider V. Voorzaat was noted in the 350cc class. The 350cc international race saw Theo Louwes 3rd with Leo Bovee 6th. In the 500cc race at Zandvoort A. Mol brought a Linto to 3rd place. Theo Louwes won the 350cc at Etten and Leo Bovee was fifth. In Venray-IJsselstein Leo Bovee gained second place in the 350cc class - his best result was at Vessem where he won the 350cc class and second place on an Aermacchi in the 500cc class.
In the world of commerce, many changes had taken place. The former importer Nimag now specialised in machines from Japan and the time of visiting potential customers with a motorcycle on a trailer had gone. Sales were now in large numbers. The new Aermacchi importer was het Tweewieler Trefpunt of Amsterdam. Their first models for sale were the 350 Sprint and the 350 TV, these being complemented by the 125cc Ala d'Oro racer. They didn't have much success up to the end of 1972, with only about ten 350s being sold. During the AMF era several hundred 350 SS and SX were sold, together with several old models which were bought back from Scandinavia. The later two-stroke models were imported and for the next ten years both new and old models were sold.
By 1971 most road racing competitors had switched to two-stroke machines. But Leo Bovee remained loyal to the 350 Ala d'Oro, gaining 3rd, 4th and 5th places in Oirschot, Vessem and Voerendaal respectively. A Dutchman was seen on a 125 Ala d'Oro for the first time at Heeswijk. This was H. Looman and John Bestebreurtje was 2nd on a Linto at Sint Maarten. In the International races at Tubbergen, G. Dodges from Australia was 9th on a 125 Oro, Leo Bovee 10th in the 350 and Alberto Pagani won the 500 on a Linto.
During the last year of Aermacchi's separate identity, Theo Louwes was 6th at Assen in the 350 International with P. Alma coming 3rd in the 350cc National race. At Tubbergen, Theo was 6th then 5th in the 350 at Vessem and 4th at Zandvoort. At Oostwolde a well known Dutchman gained a place in the 350 class on an Aermacchi. He was 6th, just in front of P. Almo, also on an Aermacchi. This was Bert Oosterhuis. At Zandvoort the positions were reversed with Alma 4th and Oosterhuis 5th. At national level it was the end of the story, but a tied 2nd place in the 250cc and 3rd in the Grand Prix final results for Renzo Pasolini provided an honourable farewell at international level.
During recent years the Aermacchi marque has come back to life in classic events. For example, in 1994 at the Assen Classic TT, Jan Kampen, the best Dutchman, gained a 5th place, the eventual winner Glauser also riding a 350 Ala d'Oro flat single.
Demonstration races have also been held at various locations by the Aermacchi Classic Racing Team, this being organised by the Aermacchi Club.

10 The Linto Project

Naked Linto, 1983.

The November 1967 issue of 'Motorcyclist Illustrated' declared, 'More good news from Italy for private riders! The famous designer Lino Tonti (ex-Benelli, Mondial, Bianchi and Gilera, now at Moto Guzzi) aided by engineer Alcide Biotti and with financial support from enthusiastic ex-rider and motorcycle builder Umberto Premoli, has prepared a 500cc racing twin for sale to privateers at the appealing price of 1,500.000 lira (roughly £865).' Of course, the story could have also mentioned that Tonti had, during the early 1950s, been at Aermacchi as chief designer.

An interesting feature of this new mount, which was to be called the 'Linto', was its space frame, with the power unit placed underneath, whilst the top end of the engine was Aermacchi Ala d'Oro bits and pieces, intended to give known reliability, easy maintenance and availability of spares at reasonable prices.

The cylinders, heads, pistons, pushrods, rockers, valves, springs, connecting rods and the like were all stock Aermacchi components. Camshafts were not standard, the camforms were, but ran in the opposite direction. So the bore and stroke dimensions were 72 x 61mm respectively; giving displacement of 496.7cc. The 33.5 degrees inclinded valves had helical springs and diameters of 39mm inlet and 33mm exhaust. Two 30mm Dell 'Orto SSI carburettors were used, these sharing a remote, rubber-mounted float chamber from which the fuel was fed via rubber ducts which were intended to 'minimise vibration and heat transmission.'

The over-the-counter 250cc Aermacchi of the period produced 32bhp at 10,000rpm on an 11:1 compression ratio. By pairing two of them, and choosing a slightly lower compression ratio, designer Linto Tonti had succeeded in obtaining almost double the power output, 61bhp@9,800rpm, with good pulling right up from 6,000rpm claimed. This was using standard Aermacchi components and Tonti was confident that he could get another five or six horsepower 'with some work on the ports and other finer points of tune.'

Tonti employed a built-up crankshaft (actually two cranks joined by Hirth couplings), with 360 degree pins (the 180 degree layout gave too much vibration and was therefore abandoned), running on four bearings. The outer bearings were of the ball-race type and there were two inner roller bearings, between which were located the gear drive to the nearside (left) camshaft and another gear which drove a countershaft that carried the rev-counter drive and on the right the primary gear drive.

In Spring 1969, Swiss-Hungarian, Gyula Marsovszky was one of the first to buy a Linto. Marsovszky is seen here leading a pack of riders headed by Bo Granath (Husqvarna) during the 500cc Dutch TT, 26th June 1971.

Alberto Pagani, with his father Nello, was the main Linto 'works' rider. He also scored the marque's only Grand Prix victory; Imola 1969.

At the offside (right) of the crankshaft there was another gear for driving the camshaft for the offside cylinder, the two contact-breakers for the battery/coil ignition, and the double gear oil pump, one side of which fed lubricant to both ends of the crankshaft, the other providing pressure lubrication for the cams, pushrods, rockers and the like.

The light-alloy crankcase was split transversely and vertically, the front half carrying the cylinders, crankshaft, oil pump and contact breaker; the rear housing the 6-speed close ratio gearbox. Oil was contained in the lower sump, which, being integral with the crankcase, was similarly split. The dry, multi-plate clutch, with aluminium plates, ran externally on the offside, whilst the gearbox sprocket was on the nearside.

An interesting, but little-reported feature was that the engine had been designed so that a kick-starter and an alternator could be easily fitted for a roadster version, which in the event was never built. Frame weight was little more than 20lb, plus an additional 5lb for the swinging arm. Both components were of steel, but with the use of molybdenum tubing plus the use of a magnesium-alloy crankcase and glass-fibre instead of a metal fuel tank. The weight, Tonti said, 'should in due course be reduced from its existing 320lb to around 290lb'.

Alberto Pagani in action at the 1969 Senior TT. After an opening lap of 97.37mph, he broke down, an event which was to occur far too often, and was to gained the Linto a bad name as well as dissatisfied customers.

Designed by Lino Tonti, the 496.7cc Linto was intended to be a serious challenger in the 500cc class. The engine comprised Aermacchi top end components, with a one-off bottom end with coupled crankshafts. It was fast, but unreliable.

18 inch tyres were 3 inch section at the front and 3.50 at the rear, the prototype Linto sporting the new 230mm Ceriani four leading shoe front drum brake, a 200mm 2LS rear stopper, and rear suspension from the same source.

Tonti also said that a 600cc engine could be built for sidecar racers, 'The horizontal cylinder arrangement should appeal to drivers using the now almost universally used kneeler outfits.' Umberto Premoli was based in Varese, near the Aermacchi works.

The British importer, Syd Lawton, saw the Linto in Autumn 1967, 'I was impressed with the space-frame, a mass of small diameter tubes, and the neatness of the installation. But I can't see a market for a machine costing in the region of £1,000 plus (in the UK). I came away admiring the ingenuity of Lino Tonti, but reluctantly decided not to purchase because of the high cost.' As things were to turn out I'm sure my old friend Syd would have been extremely pleased he didn't get involved.

Alberto Pagani and fellow Aermacchi works rider Gilberto Milani tried the original prototype machine late 1967 and early 1968. Pagani was in effect chosen to be the Linto works development pilot.

Interviewed in Spring 1969 Pagani said, ' We have improved the camshaft and the gearbox is stronger but the most important modification is better value

Tony Osborne's 500 Linto entered by Rugg's of Woodford, Essex, Brands Hatch early 1970s.

material.' Up until he finished runner up in the 1968 East German GP at Sachsenring, Pagani had been plagued by a string of valve breakages, and the resultant damage was extensive. Before the Czech GP, at Brno, the cause remained a mystery. Pagani again, ' I grabbed the clutch as soon as I felt a valve drop in on a tight corner at Brno. When I stripped the engine, I found a valve had broken at the top instead of at the bottom, as usually happens with Aermacchis.' The mystery was solved when Pagani returned to Italy. The valve suppliers had, he said, used the wrong type of steel for both inlet and exhaust valves, which (as already stated) were identical to Aermacchi components. Better quality steel ended 'the tiresome trouble'.

However, faulty material was blamed for an entirely different problem, which surfaced when Jack Findlay had a high-speed accident when testing the Linto at Monza in October 1968. The Hirth coupling failed between the two crankshafts. At the time, this was stated to have been a one-off.

Then, at the beginning of 1969 came what was generally seen ('Motor Cycle News') as the 'Road Racing Revolution'. This was in response to news that Linto would be selling a batch of machines to customers for 950,000 lire (£1,300) each. 'MCN' described the Linto as, 'a bike which seems destined to oust the old-fashioned British singles from Grand Prix racing. We hope to see at least 25 Lintos in circulation this season.' Pagani was quoted as saying 'We will complete the first dozen by the end of April.'

At that time, (16th April, 1969) seven Lintos were in existence. Six made an impressive debut at Le Mans, where production versions shattered the lap and race records of an MV four and finished first and second. The seventh, raced by Pagani at Riccione the same day, was third to Agostini (MV) and Hailwood (Honda) until the rev counter broke loose. The two works machines, used by Pagani and Findlay were painted blue and had electron crankcases, which saved 13lb.

Ian Griffiths at Brands Hatch, 1983,

How fast was the Linto? Well, Pagani claimed, 'According to unofficial timing on the fast Francorchamps downhill straight of last year's (1968) Belgian GP, I was doing 160.8mph (259kph)!'

The 1969 production bike (with 35mm carburettors) put out 64bhp. One of the first paying customers was Continental Circus rider Maurice Hawthorne. He allowed Martin Hodder of 'MCN' to test his brand new, untried bike at Snetterton in Spring 1969. Hodder seemed very impressed with the Linto, rounding off his test by saying, 'The Linto would seem an ideal machine for the ambitious privateer, not only is it faster than everything but out-and-out works machinery, but the extensive use of Aermacchi components solves the often difficult and expensive spares problem. The rider with £1,300 to invest (somewhere in the region of £25,000 in today's money) could not do better than the Linto which will figure prominently in international race results for quite some time.'

But of course the above was written without the knowledge of hindsight. If only it had been reliable... sadly things were just the reverse, and although Pagani went on to win the 1969 Italian Grand Prix at Imola, the production version proved a terrible disappointment for all concerned. Soon, there was a host of Lintos sporting 'For Sale' signs in race paddocks all around Europe.

Pagani continued to race the works models until mid 1971, before switching to MV, as Agostini's team mate, but the problems were never fully sorted.

A switch was made to Dansi electronic ignition for the 1970 season. But that Imola victory was to remain the model's only GP victory. In the end, it was the appalling unreliability of the production bikes and the rise of the big two-strokes, such as the production-based Kawasaki triple and Suzuki twin, which finally killed off the Linto project.

Dick Linton's Linto, Snetterton, 1984.

11 Two-stroke Racers

Factory rider Alberto Pagani in action in July 1967 at the Zingonia circuit, near Bergamo. This prototype was designed by German engineer Peter Durr, it produced 20bhp at 9,200rpm and had an iron barrel and 4-speeds.

Although they may not have had the character of the long-running ohv horizontal singles with which Aermacchi had entered the racing arena during the late 1950s, the new breed of two-stroke tarmac racers from the Varese factory which debuted almost a decade later did achieve Grand Prix and World Championship success, something the earlier bikes had largely failed to deliver. The original bike was a '125' single with vertical cylinder and a pretty basic specification.

The 1970 version of the Aermacchi Harley-Davidson two-stroke single. With improved porting, 28mm carburettor and 5-speed transmission, it put out 22bhp and was good for 110mph. Dry weight was a mere 80kg (176lb).

Derived from the new-for-1967 M125 Rapido, it featured a tuned version of the same 123.15cc (56 x 50mm) piston-ported engine designed by the German engineer, Peter Durr. The original prototype put out 20bhp, exactly double that of its roadster counterpart.

Intensive testing at Monza in June 1967 with Alberto Pagani, proved that the prototype - still with 4-speeds - lapped faster than the best CR93 Honda dohc twin or Bultaco TSS over-the-counter racers. So its race debut came sooner than expected, on the recently-opened Zingonia circuit near Bergamo, the following month.

After a very slow start, Pagani made his way quickly through the field to finish a promising third, behind the factory disc-valve Mondial and the Montesa of the Villa brothers, and well ahead of a horde of other machines, including not only Honda and Bultaco, but Motobi and Morini. Lubrication was by petroil mixture of 20:1. There was an iron cylinder (later alloy cylinders were tested), wet multi-plate clutch, a 27mm Dell'Orto carb, and flywheel magneto ignition.

The following year saw it modified and improved quite considerably, with a specially-commissioned double-cradle tubular frame, new cylinder head and barrel, and a new 5-speed gearbox. Power was now around 22bhp at 9,200rpm compared with the roadster's 10bhp at 6,750rpm. Although the specification included Ceriani suspension and Oldani racing brakes (and from 1969 Fontana), weight was an impressively low 80kg (176lb).

In 1969, the 125 Aermacchi two-stroke gained some excellent results, even at the highest level. With riders including Kel Carruthers, Johnny Dodds, Eugenio Lazzarini and Silvano Bertarelli, it scored points in the world championship and the Italian national series. It even won a Grand Prix, when Dodds rode brilliantly to win the 1970 West German event over the fearsome Nürburgring in very bad weather.

There were a number of either experimental modifications, or privately-built specials using the 125 (listed, like the horizontal single, as the Ala d'Oro). Two of the most interesting of these were a works model raced by Kel Carruthers and Bertarelli in 1969 and a 125 disc-valve special built by Drixl for the Milan-based Mainini brothers during 1969 and 1970.

One of the very few Aermacchi 125 2-stroke racers to be brought into the UK was raced by Roger Hill during the early 1970s with some success.

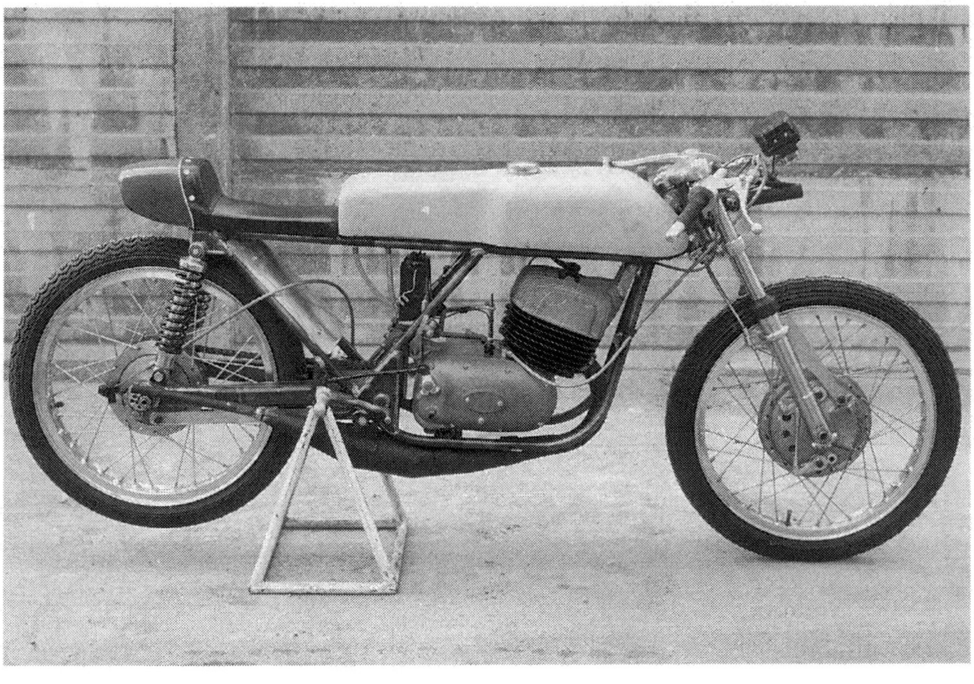

This interesting 125cc special was based on one of the Aermacchi-HD machines and constructed for the Milanese Mainini Brothers dealership, by frame specialist Othmar Drixl. Besides being 11kg (24lb) lighter, the engine had also been converted from piston-ported induction to rotary disc and a larger 32mm Dell'Orto carburettor was fitted.

1970 125 Ala d'Oro

Engine:	Air-cooled piston-port single cylinder two-stroke, alloy head, cast-iron barrel, built-up crankshaft with roller bearing big-end, vertically-split aluminium crankcases, unit construction
Bore:	56mm
Stroke:	50mm
Displacement:	123cc
Lubrication:	Petroil mix
Ignition:	6 volt, flywheel magneto
Carburettor:	Dell 'Orto SS1 28mm, with remote flat chamber
Primary drive:	Straight-cut gears
Final drive:	Chain
Gearbox:	5-speeds, foot-change, close ratio
Frame:	Duplex cradle
Front suspension:	Ceriani Grand Prix telescopic fork
Rear suspension:	Swinging arm, twin shock absorbers
Front brake:	Oldani 2LS full width magnesium hub
Rear brake:	Drum 2LS, full width alloy hub
Tyres:	Front 2.75 x 18; rear 3.00 x 18

General specifications

Dry weight:	176lb, 80kg
Maximum power:	22bhp@9,200rpm
Top speed:	115mph, 185kph

This page and opposite, Bernard Watson's, photographs, of his immaculately restored 125 Ala d'Oro, same 'Gold Wing' name as the pushrod singles.

The Carruthers' works model featured a twist-grip-operated overdrive unit bolted on to the slide of the original 5-speed gearbox. Like the earlier 50cc Kreidler GP bike, this doubled up the available ratios, in Aermacchi's case making a total of ten. The Mainini special in contrast was a complete bike, manufactured by frame specialist Othmar (Marly) Drixl. Weight had been cut to a mere 74kg (163lb), 11kg (25lb) less than the standard 1970 model. The engine had also been

Renzo Pasolini giving the new 250 air-cooled twin cylinder 2-stroke its first airing in March 1971.

converted from the piston-port induction to rotary disc valve and a larger 32mm Dell 'Orto SS1 carburettor was fitted. The frame itself was considerably lower and to match this new bodywork was designed with an extremely long and narrow tank. An electronic Krober rev counter and four leading shoe Fontana front brakes were also included in the revised specification. All in all, the Mainini 125 special was a serious attempt to improve upon the original.

'From small acorns mighty oaks do grow' goes the old phrase, and none more so than Aermacchi's racing two-strokes. The first the world saw of a brand new twin-cylinder model was when factory rider Renzo Pasolini, who had returned to the Varese factory after a four-year flirtation with Benelli, demonstrated the new machine, a 250, at the Modena Autodrome on 25 February 1971.

This newcomer was, however, not the work of Bianchi or Durr, but their successor as chief engineer at the Varese plant, Ing William Soncini. He had begun work two years before in 1969, when the American Harley-Davidson management suggested they needed a 250 racer to compete with the then new TD2 Yamaha. The engine at first used cylinders from the 125 single (which by then developed 25bhp), and therefore shared the smaller machine's over-square 56 x 50mm bore

Early 1971, the 250 twin arrived. Like the larger 350, version it was intended to replace the ageing ohv horizontal single. The engine had been designed and extensively bench-tested the previous year.

This fairing-off view of the 1971 250 twin shows the duplex frame, alloy head, cast-iron cylinder barrels, Dansi flywheel-magneto, expansion chamber exhaust, Ceriani forks and brakes, plus British Girling rear shock absorbers.

and stroke dimensions, fed by twin 30mm Dell'Ortos, coupled to flat-type rubber-mounted Amal float chambers. Electronic ignition by a Dansi (another Varese company) system powered by a generator on the nearside of the crankshaft. With a compression ratio of 12.5:1, the 248.06cc piston-port twin produced a very acceptable 46bhp at the rear wheel, revving to a maximum of 11,000rpm. Aermacchi claimed (like Yamaha) that rotary disc valves were an unnecessary complication and expense, and lubrication was also basic with simple petroil mixture. Drive to the 6-speed transmission was by straight-cut primary gears on the right-hand side of the motor, together with the rev-counter pick-up and dry multi-plate clutch assembly. The crankcase was split horizontally. Originally, dry weight was 125kg (276lb), but this was soon reduced when the heavy iron cylinder barrels were replaced by lighter alloy components, which saved 10kg (22lb).

In 1971, more time was spent on development and track-testing, rather than racing. However, Pasolini rode the new twin to fifth place at the Spanish Grand Prix; the final round of the world championship series. But there was no real hint of what was to follow in 1972 and beyond.

Originally designed in 1973, but not raced until 1975, the 498cc 4-carb twin as it appeared in 1975. Campagnolo enclosed disc front brake, rear brake disc mounted on transmission shaft and swinging arm, twin shock rear suspension. This machine is water-cooled, the 1973 prototype was air-cooled.

In 1976/77, frame specialist Bimota built a new monoshock frame for the 500 twin. This bike is the first Bimota prototype, with near horizontal shock. Later models featured a more vertical shock position and an adjustable steering column.

In racing, 1972 was to see Pasolini, the factory's sole racing representative, hurtling to Grand Prix victories in Italy, Spain and Yugoslavia on the quarter-litre two-stroke twin against an army of Yamahas. Entirely new that year was a larger version of the basic formula with a 347.4cc (64 x 54mm) engine. The larger motor made its Grand Prix debut at the West German round in April at the Nürburgring, where Pasolini finished 5th. This was followed up by a pair of 3rd positions (France and Austria) and then runner-up spot behind the formidable combination of Agostini and MV Agusta, at the home round, staged at Imola. Other second places were recorded in East Germany, Czechoslovakia and Spain. Even so, the 350 never quite matched the 250 in its class.

Pasolini got improved machines for the 1973 season, not only with more power, but water-cooling for the first time. He also acquired two new team members, Frenchman Michel Rougerie, and Bruno Kneubuhler from Switzerland. In addition, both the 250 and 350 two-strokes were offered as over-the-counter racers for the first time, under the prefix RR; the old pushrod flat singles having been phased out of production to make way for the new models.

The groundwork had been done, and early tests were highly promising, at least of the genuine works machines; the production racer replicas for the public just weren't on a par with their Yamaha rivals. Unfortunately, Lady Luck was about to deal the Varese factory a fatal hand.

20th May 1973 will go down in racing history as 'Black Sunday'. Both Pasolini and Jarno Saarinen were fatally injured in an horrific eight-bike accident which left the motorcycling world in a state of deep shock. The event was the 250cc Italian Grand Prix, at Monza, just outside Milan. The crash, which at first was blamed on Pasolini's bike, rumoured to have seized, subsequently was found to have been caused by oil left on the curcuit from the previous 350cc race.

Test rider putting an early drum-brake, liquid-cooled RR250 production version of the latest Pasolini works model through its paces in 1973. Most of these bikes were sent to the States.

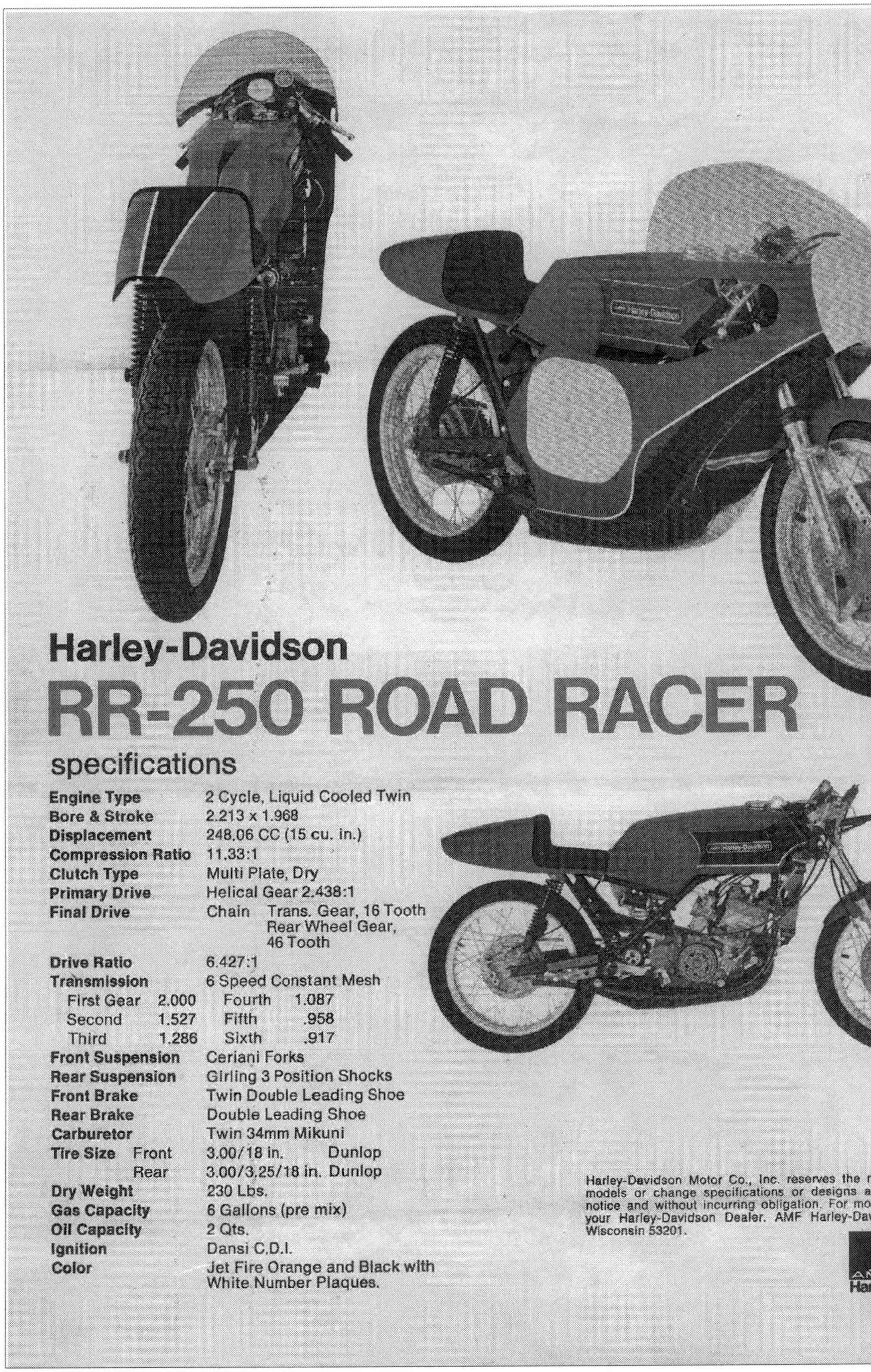

Harley-Davidson

RR-250 ROAD RACER

specifications

Engine Type	2 Cycle, Liquid Cooled Twin
Bore & Stroke	2.213 x 1.968
Displacement	248.06 CC (15 cu. in.)
Compression Ratio	11.33:1
Clutch Type	Multi Plate, Dry
Primary Drive	Helical Gear 2.438:1
Final Drive	Chain Trans. Gear, 16 Tooth Rear Wheel Gear, 46 Tooth
Drive Ratio	6.427:1
Transmission	6 Speed Constant Mesh
First Gear 2.000	Fourth 1.087
Second 1.527	Fifth .958
Third 1.286	Sixth .917
Front Suspension	Ceriani Forks
Rear Suspension	Girling 3 Position Shocks
Front Brake	Twin Double Leading Shoe
Rear Brake	Double Leading Shoe
Carburetor	Twin 34mm Mikuni
Tire Size Front	3.00/18 in. Dunlop
Rear	3.00/3.25/18 in. Dunlop
Dry Weight	230 Lbs.
Gas Capacity	6 Gallons (pre mix)
Oil Capacity	2 Qts.
Ignition	Dansi C.D.I.
Color	Jet Fire Orange and Black with White Number Plaques.

Harley-Davidson Motor Co., Inc. reserves the right to discontinue models or change specifications or designs at any time without notice and without incurring obligation. For more information, see your Harley-Davidson Dealer. AMF Harley-Davidson, Milwaukee, Wisconsin 53201.

AMF Harley-Davidson

1974 RR250

Engine:	Liquid cooled, piston-port parallel twin two-stroke, rubber-mounted
Bore:	56mm
Stroke:	50mm
Displacement:	248.06cc
Comp ratio:	11.3:1
Lubrication:	Petroil mix
Ignition:	Dansi CDI magneto
Primary drive:	Straight cut gears
Final drive:	Chain, 16-tooth gearbox sprocket, 46-tooth rear sprocket
Frame:	Chrome-moly tubing, duplex
Front suspension:	Cerani telescopic front forks
Rear suspension:	Box-section, twin Girling shocks, eccentric chain, adjustment in swinging arm pivot
Front brake:	Twin hydraulic 230mm Brembo disc
Rear brake:	Fontana 2LS 180mm drum
Tyres: Front	3.25 x 18, rear 3.50 x 18

General specifications

Wheel base:	55 inch,1397mm
Ground clearance:	5.25 inch, 133mm
Seat height:	27.5 inch, 698mm
Fuel capacity:	4 gallon, 18 ltr
Dry weight:	230lb, 104kg
Maximum power:	53bhp@11,200rpm
Top speed:	140mph, 225kph

1973 RR250. That year both works and production models were liquid-cooled. Specification included 6-speeds, 34mm Mikuni carburettors, Dansi CDI ignition, box section swinging arm and interchangeable cylinder barrels.

For 1975 the RR250 (and the new RR350) were equipped with twin Scarab calipers and cast-iron brake discs at the front. Some machines, as can be seen above, had Dell'Orto carburettors and gave an additional 3bhp.

It was ironic that Pasolini should die in such a tragic manner after one of the best races of his life. The 350cc event had seen 'Paso' come from nowhere in the first lap to challenge and then pass Agostini (MV) to take the lead. There then followed a battle royal which only ended on the 21st of 24 laps when Pasolini's bike seized; but not before he had set a new lap record at 124.27mph.

This appalling tragedy resulted in the Italian Motorcycle Grand Prix for the next eight years being staged at either Imola, Misano or Mugello - but not Monza. Quite rightly, it was felt Monza simply didn't deserve the honour, its officials had ignored warnings from riders who had reported oil all around the circuit from a leak on Walter Villa's ailing 350 Benelli four-stroke four. Not only this, but after the accident the 250 race was allowed to continue for a full two laps. One eye witness described the scene 'like a picture of hell' with bikes which had caught fire lying on the course; whilst those still circulating had to pick a path among the debris. It took more than an hour after the accident before the official news of Saarinen's and Pasolini's death was announced; once again the officials did not come out with any credit.

For the remainder of 1973, the factory provided machines on a semi-works basis for the Frenchman, Michel Rougerie; whilst in Italy they were raced on a similar basis by Gianfranco Bonera.

A single large radiator sat across the frame in front of the engine.

Pasolini's death also had the effect of halting the development of an interesting 498.8cc (66.7 x 70mm) racing twin. Although work on this potentially exciting machine did restart the following year, there just didn't seem to be the same level of enthusiasm. On the bench the big twin gave a 90bhp reading; and this excellent figure, combined with low weight, promised much. Various carburation systems and reed valve units were tested; but the real stumbling block was the problem of finding a suitable rider along with the distracting success gained by the quarter-litre twin and the need to develop a new frame design. What became of this project is related later.

There was also a 4-cylinder two-stroke 500 under consideration which was shelved as the twin seemed to offer a better value-for-money option.

Another, albeit later, development project was a 3-cylinder 350. This engine was similar to the twins, but with an extra cylinder on the nearside. In an unique arrangement, the cylinders were not all of the same size; the outer two having a bore of 56mm to give capacities of 125cc, whilst the centre one had a smaller (100cc) capacity, all three featuring a common 50mm

Top, Ex-Norton tester and rider Norman White RR350 at Brands Hatch, 6th March 1976. Bottom, Australian, Kel Carruthers with his works 125cc two-stroke single, 1969 Isle of Man TT.

RR250 at the Stafford Classic Mechanics Show, October, 2004.

Number one plate for the number one 250cc machine and rider combination in 1974, 1975 and 1976, Walter Villa and Harley-Davidson. Walter is seen here at Imola in 1977.

stroke. The three barrels and heads were all separate and water-cooled. Power output under bench test conditions was 76bhp.

In 1974 the Harley-Davidson Italia team came back with a vengeance. There was an improved 250, which was not only exceptionally fast but also very reliable! On the riding front, Bonera had been signed by MV Agusta, and his replacement was Walter Villa, who initially was intended as the 'back-up' for Rougerie. Actually, it

Frenchman Michel Rougerie, number two in the works team.

Walter Villa in action at the 1976 West German Grand Prix at Hockenheim. He won from team-mate Michel Rougerie and Yamaha star Victor Palomo.

1972 RR350 works model, as raced by Renzo Pasolini.

soon became evident that it was the Italian veteran, rather than the Frenchman, who was calling all the shots; and not just in the team, but the world. The result was the Varese factory's first ever world championship title.

Born in August 1943, Villa had begun racing in 1965. He had previously raced works machines for Benelli and MV Agusta. Villa became the 1974 world champion with victories in the Italian, Dutch, Finnish and Czech rounds; and runner-up spot in Sweden. This was a truly brilliant achievement. It was a highly competitive class with strong and unified opposition from Yamaha. By now, the quarter-litre twin was producing 58bhp at 12,000rpm, whilst the less successful 350 gave 70bhp at 11,400rpm.

Villa's championship came at a good time, as a record number of standard production street bikes - some 45,000 in total - were manufactured by the Italian factory that year. Once again, the majority was exported to the USA; but a new company, AMF International, with headquarters in London, was established in an attempt to boost European sales.

The author's former dealership, Mick Walker Motorcycles of Wisbech, Cambridgeshire was one of the very first of the new AMF dealers to be appointed; selling both the Italian and American models - plus the RR250 and RR350 road racers, which were replicas of the works models campaigned by Walter Villa at the time.

Michel Rougerie (14) and Walter Villa (2), at the start of the 1975 250cc Belgian Grand Prix, Spa-Francorchamps.

Four times world champion Walter Villa, in one of his final races on the Varese-built twin cylinder 2-stroke models, the West German GP at the Nürburgring, 1978.

Mick Walker Motorcycles' rider Mike James leads the field on his RR250 at Mallory Park, 1978.

Just to prove his 1974 success wasn't a fluke, Villa gained a second consecutive 250 world title in 1975 and generally dominated the class to an even greater extent. This was in spite of a full factory Yamaha challenge spearheaded by Venezuelan Johnny Cecotto. This time Villa won in Spain, Germany, Italy, Holland, and Sweden. Team-mate Rougerie won in Finland and Czechoslovakia, making it virtually a clean sweep for Varese.

1975 finally saw the 500 twin raced for the first time. Although fast, this four-carburettor, 90bhp projectile was destined never to reach full potential. Quite

The most successful British competitor on a production RR250 model was Mike James (25), passing a TZ Yamaha at Snetterton on his Mick Walker-backed, Tommy Stoner-tuned bike, 1977.

250cc Czech GP, 1977. Walter Villa (1) leads Tom Herron (Yamaha, 5), Kork Ballington (Kawasaki, 15) and Alan North (Yamaha, 22).

Walter Villa began racing on a Morini Settebello in 1964. The younger brother of Francesco Villa, Walter raced a wide range of bikes, including FB Mondial, Montesa, 4-cylinder Benelli, 3-cylinder Triumph and 3-cylinder Kawasaki. After the death of Renzo Pasolini in May 1973, Villa was signed up and won four world titles for the Varese factory.

simply, Villa and Rougerie already had enough on their plates with the 250 and 350 models, and the factory didn't have enough time or financial muscle to cater for a third rider at world championship level.

Eventually the under-developed 500 motor was given to Bimota who built frames for the smaller twins, and a Bimota-HD was subsequently raced in Italian events in 1976 with Vanes Francini in the saddle.

The RR250 and RR350 customer racers were by now homologated to race in American AMA events, but although the factory-backed riders, including Gary Scott and Jay Springsteen, rode them, they didn't win any titles. As for privateers, they very rarely showed, although in Britain the combination of Mike James and an RR250 tuned by wheelchair-bound Tommy Stoner did well at national level during 1977 and 1978.

1976 saw Franco Bonera rejoining the Varese team after two seasons with MV Agusta. With Villa, he rode the 'unbeatable' 250, plus the by now highly competitive 350. The year turned out to be the best in the company's history, Villa becoming double world champion.

By 1977 the shine was wearing thin. For a start, Villa was forced to defend his titles with a privately-organized team backed by the Nolan helmet company. If this suggests financial problems at the factory, you are dead right. Strapped for

Mike James on the Mick Walker Motorcycles', RR250, Cadwell Park, September 1977.

cash, Varese couldn't afford the luxury of running a proper team; the result being farmed-out bikes with performances deteriorating as the season developed. Even so, Villa's new team-mate, Franco Uncini managed to finish as runner-up in the 250cc world championship, while Villa eventually came third. The 350, however, was by now completely out of contention.

If factory rather than private support had been available, 1978 would probably have seen a new disc-valve 250 with monoshock chassis, and finally the race debut of the 3-cylinder 350 described earlier. But in fact, this never happened.

Final development of the RR250 came in 1978/79 when Cagiva took over the Varese works. The engine was revamped and put into a new monoshock frame.

12 End of the Road

Very early SX350 trial bike with one piece rocker cover, built at the very start of the AMF Harley-Davidson era.

As far as motorcycles were concerned, 1972 saw the end of the road for the Aermacchi brand name. AMF (American Machine & Foundry) acquired the remaining shareholding from the Italian company, AMF themselves having purchased Harley-Davidson in 1968. Thus, with 100 per cent ownership, like the stateside v-twins, future Varese-built bikes carried the 'AMF Harley-Davidson' logo. Some months prior to the buyout, expressly for the American market Aermacchi had carried out a redesign of its 344cc ohv horizontal roadster series (the 250 street bikes having been discontinued in 1970). This resulted in a brand new SX350 trail bike and a revised SS350 roadster.

At first, both machines employed the traditional Aermacchi-type open frame, 4-speed gearbox and the fully enclosed rocker gear with its massive one-piece cover. However, from the end of 1972 both bikes not only reverted to the conventional separate, smaller rocker covers, but also a new full duplex cradle frame and a 5-speed gearbox. Additionally, the 1972 model year SX/SS had 6-volt

battery/coil ignition, whereas the 1973 onwards version came with a 12 volt solid state ignition system and an electric start which coincided with the arrival of the duplex frame.

The frame change and electric start made a significant difference in weight, whilst the additional gear ratio helped to regain the performance lost through the first two alterations, as did a hike in compression ratio from 9:1 to 9.3:1. The 1972 SX weighed in at 309lb against 353.9lb for the 1973 machine; the 1972 SS difference being 325lb against 355lb.

Yet another notable difference was the use of a twin leading shoe front brake on the SS and a single leader on the SX. However, both used the same full-width aluminium hubs front and rear. Also the SS employed the low-level branched exhaust header pipe and twin Lafranconi silencers (the exhaust system being chrome-plated) where as SX had an upswept single exhaust pipe which ran into a massive silencer just below a seat on the offside of the bike. Both the pipe and silencer were finished in black, but there was an extensive chromed-wire type head shield on the SX's silencer. Again, the SS had a low close-fit mudguard, Whilst the SX's was high to provide plenty of clearance off-road. Both sets of guards were manufactured in stainless steel.

Harley-Davidson was bought out by the giant AMF (American Machine & Foundry) corporation in the late 1960s. One of the first bikes to be badged AMF Harley-Davidson was the SX350 trail bike, with full loop frame 12 volt electrics and electric start.

Sue Tully of Greenford, Middlesex with her 1973 SS350, mid 1980s.

To compensate for power loss due to increased weight, the SS/SX put out an additional 2bhp from the 1973 model year, making 27 instead of 25bhp. Because of the new frame introduced for the 1973 season, most of the dimensions were different from 1972 bikes. These are listed below. As for tyre sizes, both the SX and SS employed 19 inch front and 18 rear. However the section was different between the two: SX front 3.50, rear 4.00; SS front 3.25, rear 3.50.

SX 1972 (1973 in brackets)
Wheelbase: 56.3 inch (56.1 inch)
Overall length: 86.3 inch (85.8 inch)
Overall width: 35 inch (33.8 inch)
Overall height: 44.2 inch (43.3 inch)
Seat height: 31.3 inch (31.8 inch)
Road clearance: 8.5 inch (9.2 inch)

SS 1972 (1973 in brackets)
Wheelbase: 53.7 inch (56.1 inch)
Overall length: 82.7 inch (86.6 inch)
Overall width: 32.7 inch (32.7 inch)
Overall height: 41.1 inch (42.5 inch)
Seat height: 32.3 inch (31.5 inch)
Road clearance: 6.9 inch (7.5 inch)

It is also worth noting that from 1973 with the adoption of 12 volt electrics, both the SS and SX gained direction radiators. Also from the same year a switch was made from right to left hand gear-change operation. Both models had side stands only.

AMF realised that besides its American base, it needed to mount a serious sales push in Europe now that AMF International was established in Switzerland. This operated separately from the manufacturing facilities in Varese. In late Spring 1974, a British headquarters was established in London, with a spares warehouse in Whitstable, Kent which doubled up as the storage depot for AMF's bowling alley operation.

So for the first time, there was a full dealer network set up for Varese's products to concentrate upon the road bikes, rather than racers.

1973 SX350, official Harley-Davidson press photograph. Specification included stainless steel mudguards, full width alloy brake hubs, hi-level black exhaust and knobbly tyres.

Model chart 1946 - 1978

Macchi	3.MB1 3-wheel	1946-
		(3-wheel truck)
125cc	Type N	1951-1955
125cc	Type U	1952-1956
125cc	Type M	1952-1956
250cc	Bicilindrica	1953-1956
125cc	MB 8 3-wheel	1954-1957
		(lightweight 3-wheel truck)
125cc	Zeffiro	1955-1960
150cc	Zeffiro	1955-1960
150cc	Corsaro	1955-1957
175cc	Chimera	1956-1960
175cc	Ala d'Oro	1957-1965
175cc	Ala Rossa	1957-1966
175cc	Ala Bianca	1957-1965
250cc	Chimera	1957-1965
250cc	Ala Azzurra	1957-1967
150cc	Cavaliere	1957-1960
		(lightweight 3-wheel truck)
250cc	Ala Verde	1959-1970
250cc	Motocross	1959-1965
250cc	Ala d'Oro	1959-1972
250cc	Wisconsin	1961-1967
250cc	Sprint	1963-1968
50cc	Zeffiretto	1963-1967
50cc	Zeffiretto Sport	1963-1967
150cc	Brezza	1963-1968
350cc	Ala d'Oro	1964-1972
175cc	Ala Rossa GT	1966-1967
250cc	Ala Blu	1967-1970
250cc	Ala Blu GT	1967-1970
125cc	Ala d'Oro	1967-1972
125cc	Aletta	1967-1971
		(Rapido in the USA)
125cc	Aletta Scrambler	1968-1971
350cc	GTS	1969-1972
350cc	TV	1970-1972
100cc	Baja	1970-1972
125cc	R/C	1971-1972
65cc	M65 Leggero	1967-1972

Harley-Davidson 1973 and later

350cc	SS350
350cc	SX350
125cc	RC125
90cc	X90
90cc	Z90
125cc	SX125
100cc	SR100
175cc	SX175
125cc	SXT125
125cc	SS125
175cc	SS175
250cc	SS250
250cc	SX250
250cc	SST250
350cc	SX250
350cc	SST350

Competition models

MX250	1976-1978
RR250	1974-1977
RR350	1974-1977

1974 SS350

Engine:	Air-cooled ohv single with horizontal cylinder, alloy head and cast-iron barrel, built-up crankshaft with roller bearing big-end, vertically-split aluminium crankcases, unit construction
Bore:	74mm
Stroke:	80mm
Displacement:	344cc
Comp ratio:	9.3:1
Lubrication:	Gear-type pump, wet sump
Ignition:	12 volt, battery/coil, electric start
Carburettor:	Dell 'Orto
Primary drive:	Gears
Final drive:	Chain
Gearbox:	5-speed, foot-change
Frame: Duplex,	Full cradle
Front suspension:	Cerani telescopic fork
Rear suspension:	Swinging arm, twin shock absorbers
Front brake:	Drum 2LS
Rear brake:	Drum SLS
Tyres:Front	3.25 x 18; Rear 3.50 x 18

General specifications

Wheel base:	56.1 inch, 14,24mm
Ground clearance:	7.5 inch, 191mm
Seat height:	31.5 inch, 750mm
Fuel capacity:	2.8 gallon, 12.75 ltr
Dry weight:	355lb, 161kg
Maximum power:	27bhp@7,000rpm
Top speed:	84mph, 135kph

The 1974 SS350, little changed from 1973. Left-hand gearchange, 12 volt electrics, push-button starting, full cradle frame, indicators and stainless steel mudguards clearly targetted the American market.

Publicity shot of the 1974 SS350 and SX350.

SS350 power unit showing dry-clutch cover, oil filler plug, points cover, metal oil pipes, tachometer drive exit points and cable, electric starter motor and VHB Dell'Orto carburettor.

1974 SX350

Engine:	Air-cooled ohv single with horizontal cylinder, alloy head and cast-iron barrel, built-up crankshaft with roller bearing big-end, vertically-split aluminium crankcases, unit construction
Bore:	74mm
Stroke:	8mm
Displacement:	344cc
Comp ratio:	9.3:1
Lubrication:	Gear-type pump, wet sump
Ignition:	12 volt, battery/coil, Denso electric start with chain operation
Carburettor:	Dell 'Orto
Primary drive:	Gears
Final drive:	Chain
Gearbox:	5-speed, foot-change
Frame:	Duplex, full cradle
Front suspension:	Ceriani telescopic fork
Rear suspension:	Swinging arm, twin shock absorbers
Front brake:	Drum SLS,180mm full-width alloy hub
Rear brake:	Drum SLS, 180mm full-width alloy hub
Tyres:	Front 3.50 x 19, Rear 3400 x 18

General specifications

Wheel base:	56.1 inch, 1,424mm
Ground clearance:	9.2 inch, 233mm
Seat height:	31.8 inch, 763mm
Fuel capacity:	2.8 gallon, 12.75ltr
Dry weight:	353.9lb, 160kg
Maximum power:	27bhp@7,000rpm
Top speed:	82mph, 13kph

The 1973/74 SX350 trail bike was closely related to the SS350 of the same era. Besides handlebars, exhaust, tyres, insrumentation, the SX used a SLS instead of a 2LS front brake.

'MotorCycle' dated 29th June 1974 announced, 'Harleys in Britain!' It went on to say, 'The first of the new Harley-Davidsons to emerge from the Italian factory, formally Aermacchi, is a stylish 250cc two-stroke single, the SX250, first examples of which are just reaching the country.'

The new UK AMF International operation appointed seven dealers, F. W. Warr Ltd (Central and West London), Pratt's Vintage Motorcycle Co. (South East London), Park Motorcycles (North London), Peter's of Luton (Bedfordshire and Northamptonshire), Dowson's of Scarborough (North East England), Knotts Ltd of Manchester (North West England) and my own company Mick Walker Motorcycles of Wisbech, Cambridgeshire (East Anglia).

The full model range, including American v-twins, contsisted of; X90, Z90, SX125, SX175, SX250, SS350, XLCH1000, FX1200 Super Glide and the 'Big Daddy', the 1,200cc Electra Glide.

Hot on the heels of the British Launch came the first road test of one of the new bikes, an SS350, for the 6th July 1974 issue of 'MotorCycle'. Headlined 'Raw Masculinity...' it asked, 'Remember those little red Aermacchi flat singles of about twelve years ago? Italian as Chianti, their immaculate engines, paintwork and superb handling conjured the atmosphere of the racing paddock as few bikes could. They just had to be raced, and very successfully they were, too, until marauding hordes of Japanese two-strokes swallowed them up in the power struggle, but not until Alan Barnett hustled the 350cc version around the Island to second place in the 1970 Junior TT (behind Agostini's MV) at 99.32mph. The Aermacchi may have been a good handler, but it just wasn't fast enough. Yet even though that was the swan song of the marque, as a racer it lived on. The Americans always loved their four-stroke singles and the acquisition of the factory by Harley-Davidson in the early sixties provided the vital link.'

Bringing the story up to date, 'MotorCycle's' tester, John Nutting, continued, 'Now the Varese Aermacchi plant is totally controlled from Milwaukee, it seem that the 350cc singles that are being imported here in quantity have acquired the unmistakable stamp of the parent Harley factory.'

Although Nutting thought the engine itself very similar to the old, he considered the complete machine a vastly different animal from those original sportsters.

'The name Harley-Davidson has almost become so synonymous with hip, classy, but ponderous bikes that it comes as no surprise that the SS350 is a heavyweight too.' On the chassis Nutting summed up, 'Gone is that neat spine frame, replaced by a duplex multi-tube design that looks as if it could support a truck engine. The motor, still with its beautiful castings, hangs lost in the vast expanse of tubework, as if longing to be replaced with a vee-twin.'

However, Nutting felt, 'The Harley-Davidson's appearance could never detract

1974 X90

Engine:	Air-cooled , piston-port single cylinder, two-stroke, alloy head and cast-iron barrel, built-up crankshaft with roller bearing big-end, vertically-split aluminium crankcases, unit construction
Bore:	48mm
Stroke:	50mm
Displacement:	90.43cc
Comp ratio:	9.2:1
Lubrication:	Pump with separate oil tank
Ignition:	6-volt, flywheel magneto
Carburettor:	Dell 'Orto
Primary drive:	Helical gears, 15 x 48 tooth
Final drive:	Chain, 12 tooth gearbox sprocket, 30 tooth rear wheel sprocket
Gearbox:	4-speed, foot change
Frame:	Spine type
Front suspension:	Telescopic fork
Rear suspension:	Swinging arm, twin shock absorbers
Front brake:	Drum SLS
Rear brake:	Drum SLS
Tyres:	3.00 x 10 front and rear

General specifications

Wheel base:	40.7 inch, 1,033mm
Ground clearance:	5.9 inch, 150mm
Seat height:	26.8 inch, 680mm
Fuel capacity:	1.8 gallon, 8.1ltr
Dry weight:	158lb, 71.6kg
Maximum power:	10bhp
Top speed:	55mph, 88.5kph

from the bike's most endearing characteristic, its raw masculinity. Riding it, you get the impression that if ever the bike got into a head-on contest with a road roller the Harley would come out on top. That goes for the pulling power, too. From a stand-still, the bike tugs away like a train, so solid is the torque. Revving the engine to past its claimed power peak of 27bhp at 7,000rpm brought some improvement in acceleration but this was hardly necessary. Upward changes at five to six thousand were usually enough.'

Vibration was a problem at higher engine speeds and was 'most apparent through the footrests, it made cruising at 70 to 75mph particularly uncomfortable.' But 'speeds of 55 to 60mph were much happier for both bike and rider, and coupled with tarmac, tippling torque, this could be maintained virtually continuously over the hilliest terrain.'

Fuel economy was excellent, 'Even with its all-up weight of 364lb including fuel

1974 Z90

Engine:	Air-cooled piston-port single cylinder, two-stroke, alloy head and cast-iron barrel, built-up crankshaft with roller bearing big-end; vertically-split aluminium crankcases, unit construction
Bore:	48mm
Stroke:	50mm
Displacement:	90/43cc
Comp ratio:	9.2:1
Lubrication:	Pump, with separate oil tank
Ignition:	12 volt, flywheel magneto with CDI
Carburettor:	Dell 'Orto
Primary drive:	Gears
Final drive:	Chain
Gearbox:	4-speeds, foot-change
Frame:	Spine-type with duplex tubes from steering head downwards and back under engine unit
Front suspension:	Ceriani telescopic forks
Rear suspension:	Swinging arm, twin shock-absorbers
Front brake:	Drum full-width alloy, SLS
Rear brake:	Drum full-width alloy, SLS
Tyres:	Front 2.50 x 17, 3.25 x 16

General specifications

Wheel base:	46.5 inch, 1,18mm
Ground clearance:	7.3 inch, 185mm
Seat height:	29.7 inch, 754mm
Fuel capacity:	2 gallon, 9 ltr
Dry weight:	183.5lb
Maximum power:	10bhp
Top speed:	60mph, 96kph

and oil, the Harley can return 75mpg or more. At worst, 50mpg can be recorded as Martin Christie found on a flat-out run to the West Country and back.'

Starting was also praised, 'Firing up, using the electric starter mounted above the crankcases and operated by a big red button next to the twist-grip, was dead simple, and, unless the battery became neglected, first time, every time.'

Although two silencers either side of the frame 'provide plenty of muffing for the lazy beat of the engine,' Nutting considered the 'raucous tappet chatter' to be 'very obtrusive'.

As regards handling, the tester reported, 'At 56 inches the wheelbase is very long for a bike of this power and weight. The effect on the handling is to slow down the response when cranking over for corners making it harder when laying over and lifting the machine out of bends.' This was, of course, in stark contrast to the ohv horizontal single in Aermacchi days, particularly machines such as the Ala Verde 250 sportster and the Ala d'Oro road racer.

John Nutting rounded his test report off by saying that although the SS350 was 'solid, reliable and economical' it was also, for what it was, expensive at £560, but 'you won't find that blend of Chianti and Cheeseburgers anywhere else.'

Besides the two four-stroke models, the remainder of the 1974 Varese production was exclusively two-stroke. These were the X90 Minibike and the Z90 Roadster (the latter with 12-volt electrics); both used the same 90.43cc (48 x 50mm) 4-speed engine with oil pump lubrication.

The SR100 Enduro model was for the American market only and displaced 98.12cc (50 x 50mm); 5-speed, alloy head and barrel, petroil lubrication.

The SX125 Trail bike 125.15cc (56 x 50mm) had 5-speeds, engine with oil pump lubrication, alloy head and barrel.

The SX175 Trail bike and SS175 Roadster, both used the same 242.6cc (72 x 59.6mm) 5-speed engine, with oil pump lubrication, alloy head and barrel.

The only model listed above, (plus the two ohv singles) not to be sold in the UK was the USA-only SR100. This itself was a development of the earlier Baja - winning MSR100 of 1971-1972. In my personal experience of selling these machines, first as a dealer and, from 1977, as the official UK importer, the 90cc two-strokes and 350 ohv models were far less prone to problems (including

SS125 single cylinder street bike, 1975-1977. A typical AMF Harley-Davidson two-stroke.

1975 SS175/250

250 in brackets where different

Engine:	Air-cooled piston-port single cylinder two-stroke alloy head with chrome plated bore, vertically-split aluminium crankcases, unit construction
Bore:	61mm (72mm)
Stroke:	59.6mm
Displacement:	174.1cc (242.6cc)
Comp ratio:	10.6:1 (10.3:1)
Lubrication:	Pump with separate tank
Ignition:	12 volt, Dansi flywheel magneto, with magnetic pick up
Carburettor:	Dell 'Orto
Primary drive:	Gears
Final drive:	Chain
Gearbox:	5-speed
Frame:	Duplex full cradle, top tube acts as oil reservoir
Front suspension:	Ceriani telescopic fork
Rear suspension:	Swinging arm, twin shock absorbers
Front brake:	Full width alloy hub, SLS
Rear brake:	Conical steel hub, SLS
Tyres:	Front 3.25; Rear 4.00 x 18

General specifications

Wheel base:	55.9 inch, 1,420mm
Ground clearance:	6.7 inch, 170mm
Seat height:	31.5 inch, 800mm
Fuel capacity:	2.25 gallon, 11.3ltr
Dry weight:	252lb, 114kg
Maximum power:	20bhp (25bhp)
Top speed:	69mph, 111kph (80mph, 128.7kph)

warranty claims), than the 125, 175 and 250cc two-strokes. But, in fairness many of the problems were due to inexperienced owners and lack of general maintenance. One problem was that owners tended to run out of oil. Starved of lubricant, the engine would seize; the 125, 175 and 250cc single cylinder piston-port two-strokes having a flash-chromed aluminium cylinder.

The SX250 was probably the top seller. It was described in the 29th June 1974 issue of 'MotorCycle' as 'reasonably light for a dual purpose machine at 273lb dry.' It had a top speed in excess of 80mph. Besides features already described, the SX250 also came with gear primary drive, primary kickstart (meaning it could be started in gear, useful off-road), a rectangular backbone under the fuel tank which contained the two-stroke oil. Ignition was a Dansi-made electronic flywheel-magneto with provision for twin spark plugs in the cylinder head and a 32mm Dell'Orto PHB carburettor.

Ceriani forks were used and the rear suspension had adjusters for the multi-ratio springs. In line with the needs of off-road use, there was a 3.00 x 19 (later upped to 21 inch) front wheel, with a 3.50 x 18 at the rear. UK price in June 1974 was £499 including British taxes.

1974 was a notable one for the Varese factory. Not only did it win its first ever World Championship, the 250cc with Walter Villa, (see Chapter 11), but it was also a record for the number of series production street bikes, some 45,000 in that year. Once again, the majority was exported to the USA, but the new European AMF International, which included the recently set-up British operation, contributed to the total.

An interesting prototype which never made it to production, this 250cc single cylinder sportster in cafe racer guise was displayed at the Milan Show in November 1975.

A pair of improved 125cc models, the SS125 (Roadster) and SXT125(Trail) replaced the SX125 for 1975. These shared their styling with the 175/250 series, whereas the 1974 SX125 had been something of an oddball, with its bitty styling and dual rear wheel sprocket arrangement. The SS/SX 350 four-stroke had been discontinued, whilst the X90/Z90 continued in limited production but was axed mid-season. This left a six bike line-up; SXT/SS 125, SX/SS 175 and SX/SS 250.

Last throw of the dice. Just prior to AMF pulling the plug on the Varese built prototypes, a new 341.8cc (80 x 68mm) two-stroke in both street and trail guise. It was only after the factory had been sold to Cagiva in Autumn 1978 that these models entered series production.

Original members of the Aermacchi Harley-Davidson Motor Club at its inauguration, 21st April 1985.

The Dutch-based Aermacchi Harley-Davidson Club was formed in 1985. It caters for all four and two-stroke models produced between 1950 and 1978.

For 1976 a new model appeared, the SST250 with Brembo disc front brake and flatter handlebars, mainly for the European market.

But by 1977 the shine was rapidly wearing think. In the States, Harley-Davidson's advertising campaign for its series production two-stroke road and trial bikes suggested problems. There were offers of a portable radio, discounted trailer or a cash rebate. Why? Well with warehouses full of unsold bikes, both in the United States and Europe, things were looking desperate at AMF.

Into 1978 Varese attempted something of a comeback with new ideas and new models. This initiative was headed by American Ken Thorpe. What money there was went into development, rather than the Grand Prix race team which now had a private sponsor.

The new for 1978 SST350 road and SX350 trail were serious attempts to offer a wider range and bridge the gap left by the departed SS/SX350 four-strokes. The new larger two-strokes were both powered by the same 341.8cc (80 x 68mm)

A nicely restored SS350 suitably modified during the 1980s, with the addition of alloy rims, megaphone exhaust and much polish!

The lakeside Varese factory Summer 1978 post AMF closure but before the Cagiva buyout. Insert: interim brochure for the first Cagiva-marketed motorcycles, late 1978.

The author (left) with a member of his staff, at the Varese factory in November 1978 astride a couple of the first motorcycles produced by Cagiva.

piston-port engine, clearly developed from the existing 175 and 250cc powerplants. Whilst pre-production hand-built prototypes showed an excellent turn of speed at over 90mph, the actual production versions were a distinct disappointment. These vibrated and were prone to engine seizure, something the smaller models would never do unless starved of oil. Worst of all, they proved no quicker than the 250s! Like the 175 and 250, the oil tank for the engine was incorporated into the top frame tube.

There were also new MX motocross and RX enduro dirt racers in the pipeline; the motocrosser being a development of the original MX250, sold in the States in 1977, (see Chapter 7). In addition, there was a new 125 street bike, the SST, with Brembo disc front brake, and a revised version of the SXT125 trail bike, with improved switchgear, instrumentation and other smaller details. But this all came too late.

The fact was, AMF's corporate bosses over in America had already taken the decision to pull the plug on their European manufacturing activities. The official announcement was made in May 1978. The Varese subsidiary was to be put into liquidation in late July, when the whole of Italy traditionally closes for almost four weeks of annual holiday. During this time, an ambitious new local company, Cagiva emerged as the clear favourite to buy the plant.

At the time no-one could have predicted what was to happen. In little over three years Cagiva became Italy's largest motorcycle manufacturer, and in 1981 the SST125 (derived from the old Harley-Davidson model) was the top seller in the all-important 125cc class in the domestic market. In addition, Cagiva designed a

One of the first projects mooted by new owners Cagiva in Autumn 1978 was a couple of horizontal singles based on the old Aermacchi design. This is the touring model. There was also a sportster which featured enclosed disc brakes and Campagnolo alloy wheels. Neither passed the mock-up stage. In fact the saddles were made of wood!

brand new 125 motocross bike, which in 1980 became the world's first liquid-cooled production dirt bike racer in this popular category.

As the 1980s unfolded Cagiva began swallowing up other marques, including Moto Morini, Husqvarna, CZ and its biggest scalp of all, Ducati. It also purchased the rights to the legendary MV Agusta brand name.

Then in the 1990s everything began to go wrong, and with huge debts it was forced to sell off Ducati and cease production of both Moto Morini and CZ. However, by the end of the 1990s it did manage to re-organise and re-launch the MV marque, with its magnificent new four cylinder range.

Motorcycle production of Cagiva and MV models (under new management) continues to the present day at the old lakeside works at Varese on the very sight, all those years ago, from which the world famous Macchi seaplanes had been launched.

The 294cc Aermacchi type engine in the Varese factory after the Cagiva takeover, October 1978.

13 The Classic Re-birth

Post race debrief, Isle of Man, 1996. Chris Swallow with his back to the camera, Bill Swallow, Ron Gardner (Gardner Carburettors) and Bill Horsman.

South African Aermacchi enthusiast, Herbert Namink's superbly crafted Ala Verde-based café racer, 1985.

The overhead-valve horizontal single is to many enthusiasts the embodiment of the Aermacchi marque, in much the same way as models such as the Manx Norton, BSA Gold Star and Gilera Saturno were to their respective brand names. For all of these machines and many more, the Yamaha TD2 , 250cc and TR2, 350cc two-stroke twins in 1969 altered the sporting motorcycle as never before. Virtually overnight, long revered names simply became uncompetitive. During the early 1970s these bikes and many more were put away and forgotten.

The Classic era, which began in the late 1970s changed everything yet again. Those motorcycles which a decade ago had become redundant, were in demand once again. The Classic Racing Motorcycle Club (CRMC) was the first of many such organisations to run historic races for bikes such as the Aermacchi Ala d'Oro.

The following pages are devoted to some of the leading figures in the Aermacchi classic movement, with personal reminiscences from such respected figures as Dick Linton, The Bladon Brothers and Bill Swallow.

The Dick Linton Story

Dick at the bottom of Bray Hill, both wheels off the ground, 1983 Manx Grand Prix.

I first got the racing bug at Thruxton, Easter 1961, although I didn't get around to trying it until 1970. In between I had trained to fly, married and had two children. Then I fell for Aermacchi so it was little surprise that when I did start racing it was on a 1966 250 Aermacchi bought from Lawton and Wilson. The first race was at Snetterton, 29th March 1970, on the old long circuit. I was lying 6th (not very many more in the race if I remember correctly!) when the exhaust valve dropped. Lesson one: Measure the oil in, don't use the dipstick! So back to Syd for the first of many visits to the 'Malthouse', his Southampton premises. It wasn't the start you dream of. I had the bike until August 1971 and had little but trouble with it. Looking back through my notes, much of it was down to my own inexperience but there was a persistent problem with piston ring breakages. Syd eventually tracked it down to a faulty batch of Borgo pistons where the ring grooves were not deep enough, most unusual as the Borgo pistons were very good for the time.

So in 1971, I part-exchanged it with Syd for a dry-clutch long-stroke 350. The chassis was from an ex-works 382 model which Enzio Mascheroni (chief race mechanic) had been using as a suspension test-bed. Baroni forks were fitted and alternative top mountings for the rear units which had them laid down like early

Suzuki RG500s. Into this Syd fitted an ex Ken Kay/ Mike Trimby engine, rebuilt by son Barry. First race with this was at Crystal Palace 4th September 1971 which proved that I was no competition for the likes of Mick Andrew and Ray Pickrell. What was left of the season was spent learning how to set the bike up properly, or at least better, and to prevent the front tyre touching the cylinder head under braking.

About this time, Syd, introduced me to Francis Beart, who was also visiting him. Syd suggested that Francis look after my engine since he lived only seven miles from me. This was a turning point for me after which I spent lot of time at Brook. Mr Beart, as I knew him then, had an austere reputation. Certainly he didn't suffer fools gladly but on the other hand he did have a good, if very dry, sense of humour. If you took your racing seriously and were prepared to listen he could not do enough to help you, or so I found.

1972 then, was very much a learning year. Francis asked if I would be prepared to lend the bike to Bob Bentley for the Manx Grand Prix, provided it was returned in the same condition. This was agreed, Francis stripped the engine and Dudley Ward fitted a new big-end. Bob did approximately 875 miles on the Island so obviously reliability was coming. The only problem he had was with the Baroni forks, which had him in the ditch on the flat last right before the Black Hut. The problem was cured by fitting Ceriani fork yokes which restored the trail to the designed figure.

During 1973, against Francis's advice, I part-exchanged the long-stroke engine with Syd for an ex-works semi-short-stroke (77 x 75mm). This was the pair to Syd's Alan Barnett engine. Francis had in fact borrowed it for Jack Findlay to use in the 1970 TT, in his (Francis's) chassis. They had nothing but trouble with it. Anyway, I took it now fitted with Dunstall twin discs, to the Island, for my first Manx, The Golden Jubilee. There was not too much golden about it for me though. It misfired the whole fortnight and I finished the race, just, with rings in a hundred pieces and one pint of oil left, in last place. At least I got a finishers award!

In 1973 Ceriani had started selling their new GP disc brake forks. In common with most boy racers I became a firm follower of fashion and had to have a pair. Nobody sold them in the UK at the time, so I jumped on an aeroplane, rented a car and ended up bringing a few pairs back with me. Some went to Tony Foale from whom came a Drixton Aermacchi frame and swinging arm, plus the later Offenstadt cast alloy wheels plus discs at both ends. This was all wrapped in a Spondon-supplied Peel-style fairing. Into this the short-stroke motor was fitted and the surplus Aermacchi chassis was sold to Stuart Morrell. The result was, I still think, a very pretty bike. Some time into the 1974 season I remember reading an article by Peter Williams advocating running tubeless tyres. Save 1.25lb at the rim, improve the handling and save weight overall was the gist of the article. Come

Dick Linton, Drixton Aermacchi at the Gooseneck, 1974 Manx Grand Prix practice.

August I was off to the Island with no tubes. Practice was okay until Thursday when I had Dunlop's fit new tyres.

At Ballacraine on the first lap the bike started to wriggle. The Drixton frame was well-known for tube breakages so I had a quick look around and carried on. It seemed okay but then started again from Glentrammon through to Milntown so I toured in fourth gear, 100mph, into Ramsey. I braked to a halt in Parliament Square, turned the bars to lean the bike against a wall and the front tyre fell right off the rim: not just into the well, right off the rim! Nobody brought me a pint either! It turned out Dunlop had not yet started making non-porous tyres. A new tyre and tube were fitted free of charge.

Earlier in the year Syd had approached me to ask if I would be interested in buying the concession from him. Hampshire County Council were going to do a compulsory purchase on the 'Malthouse' for a new road, a good opportunity for Syd to go into semi-retirement. Francis thought it would be a good idea and that the Aermacchis probably had another five years in them! So I took the plunge and bought all the production spares, Syd keeping the works stuff.

At the end of the season I sold the Drixton and bought back the 382 chassis from Stuart Morrell. He had fitted an ex-Keith Dixon long-stroke engine into it. This engine had been looked after by Francis and was a known quantity, and I was fed up with problems on the short-stroke at the time. I still have the bike, though not now with the same engine. It has had two wins, seven 2nds, one 3rd and two 4ths in Classic Manx GPs. The Newmarket Motorcycle Club had by then started a single cylinder championship and this was really the forerunner of Classic racing. I managed 2nd in the championship to Ian Griffiths on the ex-Spaggiari Ducati. The most significant thing that year for me however was that I managed to buy from AMF Harley-Davidson all remaining Ala d'Oro spares, plus drawings and permission to reproduce components.

Racing in 1976 brought much of the same, 2nd again to Ian Griffiths in the Newmarket series, plus my first replica in the Manx. Business-wise, the Varese factory found five incomplete works engines and so I bought those as well. At the end of the year I decided to build a 402 model to give me a double ride at the Manx. Francis did the engine and I did the chassis, which was built with the 'scrambles' (ERS actually) swinging arm favoured by both Syd and Francis. This had lower bottom suspension mounts and permitted the use of longer units.

1977 was a typical year of short circuits, but the Manx brought a lot of hard work. The 350 ran well all through practice, but the 402 broke a gudgeon pin on the 2nd lap of practice at Ballaugh. Barry Lawton was coming over to the Island mid-week and brought a new rod and big-end from Syd, which Francis then put together. Almost immediately it broke a crankpin so it was put to the back of the garage. I decided to ride the 350 in the Senior provided it lasted the Junior. The Junior race was delayed 24 hours and the 350 broke its crankpin at Kate's Cottage on the 4th lap. I thought that was it for the week, but Alan Cathcart had fallen off his 350 near Glen Helen on the 2nd lap and he offered his engine. It wasn't a straight-forward swap. His engine had an offset carb, so I had to use his tank as well. For some reason we also had to swap the swinging arm and back wheel, which we pinched from the 402. Work started about six o'clock and someone was despatched to get an extension for the weigh in. Stuart Morrell and Ian Griffiths were both helping! I packed up about 11.00pm to get some sleep and they worked until 3.00am. I got up at six to set the carburation and run it. It was all worth it. I had a lovely ride. Les Trotter won on the Crooks Suzuki. I caught the rain over the Mountain on the last lap and just missed out on a replica. Not bad on a 350 cobbled up over night in 1977.

That was my last Manx Grand Prix proper, ie. not a classic Manx. By then the ACU had started the Formula I, II and III TT races, and with the help of the Italian factory I was able to get the Aermacchis homologated for the Formula III. I didn't go myself in 1978 but did in 1979. It was nearly my best result. At the end of the first lap Bill Smith was leading from Steve Tonkin, both on works Hondas. I was 45 seconds down on Tonkin but by the Gooseneck I was signalled second.

Ian Griffiths, Brands Hatch 1983, Linto 500 twin.

Sweeping into the Bungalow the front brake cylinder against which the cable balancer lay, snapped. The lever came straight back to the bar. I managed to slow enough, but as soon as I let go of the lever, the cable set up dismantled itself and that was my race over. I had taken 45 seconds from Tonkin from the start to the Gooseneck. Had I only held that time to the end of the lap I could have had the lap record. What's more Bill Smith had to stop for fuel and I didn't. Ah well!

1980 brought my only injury in the Island. I had taken John Wittman and his Aermacchi over and was staying with Alan Cathcart at Groudle. At the end of practice week we went to the White City Go Karts. Not all were racers but most were. Witty and Alan shot round the outside of a non-racer, I tried to dive through on the inside, he moved over and I found the last tyre on the right was full of concrete! I had a very sore left ankle x-rayed at Ramsey. No problem they said, get out and walk on it. I did and it hurt, so we borrowed a stake out of the paddock fence as a walking stick. Fortunately, the race was delayed 24 hours due to poor weather. Alan had to go home due to business commitments, which was also fortunate for me. I borrowed his left boot, which was a size larger than mine because I couldn't get mine on! On race day Dave Hook got permission for me to start last with a pusher. He took the bike through scrutineering and I avoided it like the plague. Still I finished 12th. Witty drove home to Luton and I had to drive from there with son Mark, then 12, to help, around the M25 with him operating the clutch! A week later I was in plaster, nine months off flying.

The Classic Club had started in 1979 of course, so there was renewed interest in Aermacchi. By now I was flying for a living, running the business, helping with the Classic Club and racing some as well… I was busy!

For 1981 I also had a Linto, after all it was two Aermacchis, though the workload was more than double. Alan Cathcart and I had bought all remaining spares from Premoli in Varese, and I had built it from new original parts, a proper replica.

By now, encouraged and coached by Francis and Syd, I was starting to do my own engine work as well, but it was to be some time, 1992 in fact, before I started getting more adventurous with engine development. However I had to be (more adventurous) with parts, as originals were beginning to run out. Valves, pistons, gears, etc had proved relatively straight forward, and we had managed to improve crank pin durability. I had also been able to source new old stock from as far away as Singapore. Major items like crankcases came on stream from the mid 1980s (not without problems!) as did frames, courtesy of the Hertford-based Harris concern, and with the classic boom, wheels, forks etc were becoming readily available. Things were looking up, and we had had a good five years so far!

My son Mark started racing in 1983 as soon as he turned 16, winning his first race still wearing his novice jacket on the Aermacchi. By 1985 with everything else and now preparing bikes for Mark as well, it was time for me to stop racing myself. Fortunately I got as much pleasure from running the bikes as I used to riding them. Mark carried on with the Aermacchi at Classic events and in 1986 did his first Manx, coming 4th. 1987 was DNF, 1988 2nd and 1989 3rd. By then he was riding Yamahas as well and decided he wanted to concentrate on them and drop the 'Macchi. So that was when I approached Bill Swallow for the 1990 Manx GP. It was the start of a mutually rewarding partnership, which has been a lot of fun too.

The 1990 Manx saw Bill up against brother Richard on Syd's short-stroke. No chance I thought, but Bill thought otherwise. By the Thirteenth Milestone Bill was in front, but the right-hand crankcase cracked. All Bill knew was that he had an oil leak, so each lap he went straight on at Govenors Bridge. While he turned round he had a quick check around. 'I just tried to keep it upright on the flat out bits,' he said. He finished 2nd with the case cracked through over two thirds of its width!

In 1991 he was lying 2nd at the end of the first lap, but struggled into the pits at the end of the second. I fitted a new battery and plug, and pushed him off, more in hope than expectation. He left the pits 7th and finished 2nd, enough said.

During 1992 I built a short-stroke to try and take on Syd and Richard properly. Sadly, before that could happen Richard was killed in a domestic incident. By July I was still waiting for pistons ordered months before. In desperation I called Syd and he found an original Borgo one from his scrap bin. It had broken rings stuck

Mark Linton at the Gooseneck in the 2003 Junior Classic Manx, he retired in third position on the last lap when the gearbox bearing went.

in it. We managed to get them out and clean it up and that's what was used. I had a second chassis (standard) by then, so the short-stroke went in and I took the long-stroke as a spare. Things went okay with the short-stroke so the long-stroke was not used a lot in practice. Strolling through the paddock on Thursday, Geoff Tunstall jokingly asked if I had a spare. His 7R was giving him nightmares. I had a word with Bill and Trevor White (Bill's number one mechanic for years), and both agreed a Huddersfield one-two would be nice, so Geoff qualified on it on Friday. Then I had to strip it to sort a gearbox problem. Race results Bill 1st, Geoff 2nd. Syd was delighted for me, congratulated me, then charged me £75 for the piston!

For 1993 the short-stroke was revised with a one-piece crank and a nicasil liner. The crank necessitated a revised oiling system, to ensure a positive feed to the big-end. An early (pre '64) oil pump was used. This had a pressure and scavenge side. By re-routing the internal oil-ways, both sides of the pump became pressure sides with the former scavenge side feeding only the big-end. The original pump side fed the camshaft and the rocker. To get oil from the pump to the big-end, a camshaft outrigger bearing carrier was made from solid billet with braided oil pipes taking the oil via a Yamaha OWO1 pressure relief valve to the crank feed. To convert it, a 1966 type side cover had to be used. Everything worked well enough but there appeared to be no discernable benefit.

The night before the race, with the bike all ready, Bill decided he wanted to practise some push starts somewhere quiet. He duly returned with a long face - the bike had given no trouble all week, but now the bush in the sleeve gear had picked up on the gearbox mainshaft and locked the two together. The engine had to come out, head and barrel off and the cases split. Then we had to grind the bush off, clean it all up and reassemble. Bill went back to the digs and at midnight we were out in the countryside and I was running it up to fouth gear to be sure all was well.

In the race, after a tremendous scrap with Bob Heath, Bill broke clear and with 12 miles to go, the layshaft bearing broke up and destroyed the gearbox.

The 1994 Manx was no more successful, with a cam follower failure whilst in the lead again. 1995 was little different in results, but the story is interesting. Bill had fairly controlled practice week and we were confident for the race. When we got to the bike in the warming up enclosure before the race, the cover was up at the front. We thought nothing of it, and checked the tyres etc and got Bill on his way. The bike misfired from the drop of the flag and never went at all. Coming down to the Creg the jet tube extension unscrewed from the jet tube, no more fuel to the engine but plenty to the back wheel. This thread is a taper thread and it was secured by Locktite. It is unheard of ever to break the joint, it certainly raised questions in our minds!

Bill now raised the question of building a 250 to have a go at the Suzukis in the Lightweight Manx. Given that they had to stop for fuel and we wouldn't, we felt it would be possible, if challenging. I didn't want to give up on the Junior so approached Bill Horsman to back up Bill in 1996 and go for the win in 1997 with Bill on the 250. Another short-stroke engine was built and put into my old faithful chassis. Bill won the race with lap and race records, the race record still stands in 2005, with Horsman second. At the end he had eleven spokes on one side of the back wheel broken and the hub was scrap. He thought 'it was getting a bit lively!' Horsman had a good ride after a troubled practice. Early on he ran a big-end then in Wednesday evening's practice he arrived at Ballacraine just after a shower of rain. By the time he realised he was committed! 'I knew I was going to come off,' he said, 'but I tried to get under the bike to keep it off the road.' A great character!

Over the winter of 1996/7 I set to and built the 250 engine. The spec was that of a late factory ultra short-stroke engine, 74.2 x 57.8mm bore and stroke, which in the works era ran to 12,000rpm. The Frecchia exhaust valves of the day tended to lose their heads, but we weren't using those. We were using a titanium rod though. To help things along, the chassis was put on a diet. In particular I was fortunate to locate a pair of original new Campagnolo disc hubs of 1966 vintage. With discs both ends and other lightweight bits and pieces the bike eventually

1999 Manx Grand Prix, waiting for the start Bill Swallow's 250 on the left, Bruno Leroy's 350 on the right. Dave Varney, Bruno's mechanic, behind, and Peter Berwick on the right, eye up the opposition. In the race Bill broke down and Bruno was second.

weighed 200lb (91kg) with oil but no fuel. It seemed to go quite well too. Horse, as I've said, was coming over to ride the 350 again. Practice went okay for both. Bill started well in the race and was, I think, one second in the lead at Ramsey hairpin. Then the piston broke up. Horse meanwhile had not started well on the 350, Bob Heath pulled out 30 seconds on the first lap but then Horse managed to hold the gap. Things remained fairly static until Bob stopped at Glen Helen when the chain came off. He managed to get it back on again and continued in second place. Then he got black-flagged over the mountain whilst the bike was checked over. He was allowed to continue and was credited with

Bruno Leroy in the winners enclosure , 1998. Jason Griffiths on the left and Bob Heath right.

the time. So Horse came home, the first Australian to win a Manx, at 59 years of age! Commiserations to Bob but we had had our disappointments in the past too. At the time I was surprised that Horse's speed was slower than in 1996, but I found out later. During the intervening year he had had an operation on his legs. This had gone wrong and it had to be done again. He was far from fit for the race, but we knew nothing of it, as I said, a great character. At the time there was an age limit of 60 for the race, so that was to be his last Manx.

Bill wanted to try again with the 250, so now I needed a new 350 runner. Bruno Leroy had ridden well in the 1997 Senior so I approached Patrick Godet, his 500 sponsor to see if he would be interested. He was, so all was well.

His first ride for me as at Paul Ricard, Easter 1998. Bill, John Poyner and I drove down with two 350s. Bill was also to ride John Oldfield's Triumph Triple for the first time. At this meeting the 350s and 500s ran together, and there was timed qualifying. Bill went off in his way, but Bruno was very slow, running ten seconds or so off the pace, lap after lap. I wondered why I was loaning this guy the bike. Then on the last lap he just put it on pole, in front of all the 500s. Bill was 2nd. In the race Sandro Bauman (500 Seeley) took the lead with Bruno 2nd. Bill messed up the start but worked up to 3rd. Bruno passed Sandro and cleared off, a good start. Then later he was due out on his Honda 350 in the unlimited and post-classic race with Yamaha TZ 350 and 750s, plus Kawasaki 750 Triples, Triumph Triples etc. Bill suggested he might ride the 350, which Bruno jumped at. Just before the start, a clutch start, it came on to drizzle. Bill had qualified near the front but Bruno was well back on the 7th or 8th row. Aermacchis are not noted for sweet clutches so I couldn't believe my eyes when Bruno came round 3rd at the end of the first lap. Four laps later he had dropped to 4th with Bill 5th. Then he lost it at the flat right at the end of the Mistral Straight. He lost the end of a finger but was otherwise okay, the bike was in a bit of a state though. Back to reality!

AT the Dutch TT some weeks later Bill won with Bruno 3rd, getting his confidence back. He was the first Frenchman to win a Manx, beating Bob Heath. Meanwhile on the 250 Bill was running well up when half a valve head fell off, and that was that. By now we were running a new type of rod and big-end. As I said before in the 1980s we had made the crankpins more reliable, but they were still a shouldered type and ran on an original type big-end. Now we had gone to a plain un-shouldered pin with a new steel rod and INA bearing. This has proved very reliable, and is being used in the 476 currently, giving 47bhp at the rear wheel. The steel rod now weighs less than the old titanium item.

During the winter of 1988/99, encouraged by Bill Horsman I put together a 445, essentially the same as Syd had run with Richard Swallow and Syd's son Ian. A

Tony Myers at the Gooseneck in the Senior Classic Manx, 1990.

Mechanic Trevor White, a very good racer in his own right and a great bloke, warms up Bill Swallow's 350 Ala d'Oro, for the 2002 Junior Classic Manx.

few bits were different to Syd's, but both were based on the factory 408 (78.5 x 84mm bore and stroke). The factory itself took it out to 422, which Syd also did, before taking it further to 445 (82 x 84mm). Our engines were very similar to the originals with original head, barrel, cases, crank, steel rods, N6 cam, but both with 6-speed gearboxes (Schaffleitner- Syd's original, mine a copy). The first race for mine was the 1999 TT Classic which Bill won ahead of Jason Griffiths on Fred Walmsley's Manx Norton.

The 1999 Manx gave mixed results. Bruno was second to Bob Heath in the Junior. Bill was running well on the 250 only to have the gear pedal break, same design as usual but made by a different supplier! Tony Myers, on the big'un, retired at the Gooseneck on the first lap with piston failure.

I was keen to get Bill back on the 350 for 2000 but he persuaded me to have one more go with the 250. As Hondas were now back in as well, I decided to lighten the 350 to the same spec as the 250. I also tried 415 chain from a 125 on all the bikes, which turned out to be a mistake. Two days before Manx entries closed, no entry had been received from Bruno and I was unable to contact him. So I asked Wattie Brown if he would like a ride. A rapid entry was made and then Bruno's turned up the next day so now we had three. Trevor White, Dave Varney and I were going to be busy. Wattie tried a semi-short-stroke to start with, and then an experimental ultra-short-stroke (80 x 69.5mm). This was intended to be run at 9,500rpm, but more of that later. Bill had a problem early in practice when the chain came off the 250. We put it down to high frequency vibration, all the vibration on an Aermacchi comes out at the rear of the bike due to the horizontal engine, and neither of the 350s seemed to have a problem. As we didn't have any sprockets for the 428 chain for the Campagnolo hubs I had a drum rear wheel sent over on Manx Airlines. That solved the chain problem we thought, but in the race Bruno's chain came off on the first lap, which he replaced, then it came off again on the second lap and he retired. Wattie, despite his weight, got away with it. He was quite lucky to eventually finish a good 2nd to Jason Griffiths (Honda). Throughout the race he had a big hole on the carburation, so ran it to 10,500rpm and didn't use top gear. At the post-race strip we could see the piston had been hitting the head, so that was another for the scrap bin.

As for Bill on the 250, he was having a great time. After the pit stops, not for him, he got into a great race with Roy Richardson on a Suzuki, slipstreaming him and passing under braking, only to have Roy come back on what was unquestionably a faster bike. Bill lapped at over 98mph, the fastest ever on a 250 single. He is also the fastest on 350 single at over 102mph and on a 500 single at over 108mph on Brian Richards' Norton, not a 600, as often stated. The race was so close Bill would only admit to revving 'over 11,000rpm'. Ten miles from home the top came off the piston. That was the end of the 250 project.

2001 was the year of foot and mouth, and for 2002 the line-up was Bill and Bruno again. Bill won with Bruno 4th.

During the winter I revised one of the semi-short-stroke motors with a larger rod and a more compact slipper piston, in search of more torque. Bill liked it so I converted a second engine. The first engine, when optimised on the dyno, needed one degree more of advance on the ignition than previously, so of course that is what I set on number two. More of that later. Bill was to ride the 350 in the Junior, and a new 476 in the Senior, developed from the 445. Son Mark was to ride Bruno's bike in the Junior. Practice was problematic. Bill's 350 had a persistent oil leak. When the barrel was removed to sort it, it was seen that the

Tim Stevenson, son of well know frame builder Keith, stepped in at short notice pre TT Classic when Bill was injured. Despite being a newcomer he won both 350 and 500 Singles with 2nd in the 350 Open and 2nd in the Unlimited. He is seen here at Castletown Corner.

bottom of the piston had picked up. We thought it was bore wash as we had a leaking petrol tap. The liner was unmarked so we fitted a new piston and ran it in. A new petrol tap too! In Friday's practice the engine tightened so we had to change to the number two engine, still to be run in. This would have to be done on Tuesday after the Senior race. Mark's 350 had been going well until it broke a clutch, but this was easily fixed. The 476 initially had a cylinder base oil leak, soon cured. Then on Wednesday evening it stuck in bottom gear at Laurel Bank after the back brake locked on. The engine was removed and completely stripped to reveal a bent selector fork, which was replaced. The brake was stripped, cleaned and bled and was no further trouble. The Senior was up first and whilst optimistic for the leaderboard if the bike kept going, neither Bill nor I expected to beat Alan Oversby on the Craven Manx. Alan lead from the start with Bill second, and then the Norton dropped a valve on the 2nd lap, Bill inherited the lead and soon caught Wattie Brown and John Loder who had started 20 seconds ahead and were contesting 2nd place. He stayed with them until Wattie retired when his needle clip broke. Bill stayed with John despite gearbox problems for the last two and a half laps, being unable to get below 3rd much of the time. Nonetheless, he lapped at 105.7mph on the 3rd lap. On the last lap he did get below 3rd at Ramsay Hairpin, but then got stuck in 3rd gear up to the Gooseneck. When he managed to get above 3rd, John was out of sight. He got his head down and caught up by the Bungalow, at the expense of another pair of socks, worn through to his toes! John very sportingly let him by to finish ahead on the road. The first win by a pushrod engine for many years, for which I feel Bill did not get the credit he deserved.

On Tuesday, between races, Bill ran in the 350 engine at Jurby. All seemed okay, but on race day the bike was clearly not right with a big hole in the carburation and unable to pull the gearing. He settled for 3rd behind Jason Griffiths and Chris McGahan on Hondas. Chris eventually stopped so Bill inherited 2nd for his final Manx. Mark was running a distant 3rd until Creg ny Baa, on the last lap, when a lay shaft bearing broke up, yet again. At the post race strip Bill's piston looked well rich so I was anxious to get it onto the dyno when we got home. It was 3bhp down with a huge hole in the carburation, just as Bill had said. It would not respond to carburation changes at all, and sounded as if it was detonating. After some head-scratching the ignition was retarded one degree. Instantly 3bhp back and carburation responding normally, so as you do, we took another degree off the advance and lost it all again!

Over the winter 2003/4 a lot of attention was given to sorting the gearbox problems, in particular on the 476. The 2004 season started indifferently with Bill somewhat subdued. He woke up with a vengeance at the TT classic. He won the 350 Single with new lap and race records, 3rd in Open 350, won the 500 and led the Unlimited until a trip up the slip road dropped him to 2nd behind John Loder on the 900 Weslake.

Since 2003 Bill and I have concentrated on enjoying ourselves. In 2004 he has had some fantastic rides, particularly on the 476 which he really enjoyed. I've been playing around with some variations on the 350s, most of which haven't worked but from which I've learnt a great deal. I am still trying to put it to some use. Given that we are still using original strokes, iron liners, original design pressed up cranks, albeit modern bearings, steel rods, original cams, steel valves, etc unlike the current Nortons and G50s, it says much about Aermacchis' original concepts and quality of materials that we can still be competitive.

At present (Winter 2004/5) there is a significant development programme under way on the 476 in the capable hands of ex-Cosworth technician Eddie Griffiths. When that is complete he will be taking a look at the 350 as well. Over the years I have continuously tried to refine and improve the bikes. This however promises to be a significant step forward, so hopefully we'll be able to give the latest 'replicas' and Hondas a run for their money. We shall have to wait and see.

Dick Linton.

Bill Swallow, 2003, Senior Classic Manx Grand Prix. Leading on the second lap, at a speed of 105.07mph.

Aermacchi Memories: Barry Lawton

Barry Lawton demonstrates how easily the pushrod Aermacchi racing engine could be removed from the frame, late 1960s.

I remember early Aermacchis (1962–1965) as quite agricultural looking. They steered well and the brakes were pretty good, especially early Oldani F1 assemblies, but the engines were extremely fragile. If it wasn't a disintegrating piston it would be a con-rod breaking through the crankcase and then valves dropping in (early engines had Freccia components), seats loose in the head, pushrods collapsing, crankpins breaking, wet-clutches that were very heavy to pull in and still wouldn't disengage or points floating and altering the timing. They had slim tanks and very narrow metal seats that broke.

Whilst in hospital with a broken leg (car accident) in 1962, I sketched a seat that also acted as a rear mudguard at the front. Along with fibreglass tanks, this was made by Doug Mitchenall, making the bike more comfortable to ride. But in all fairness, these racers (1962 - 1965) were just modified road bikes, originally designed in the late 50s and only having a maximum capacity of 175cc.

But for 1966 Aermacchi designed a completely new frame, swinging arm, stylish fibreglass tank and seat. Also, they fitted 35mm Ceriani front forks and wheels. Syd had a copy of a factory fairing made by Doug Mitchenhall and we bought lovely alloy and metal fairing brackets from Angelo Menani in Milan. When these were fitted, plus changing the standard chrome bars for Menani's beautiful light-weight chrome molybedenum tube clip-ons, the bike really looked like a proper racer, and with a few modifications and paintwork, you could make it look just like a factory bike racing in the Grand Prix.

The engines had been beefed up with totally new, stronger crankcases, better flywheels (with outside flywheel in 1967), stronger double-webbed con-rod, better pistons, shorter strokes with better barrels and wider finning and totally new cylinder head with bronze seats and light-weight valves, stronger springs and alloy collets, I believe some of this stuff was American.

Now we had a bike that looked like a proper race bike, went like a proper race bike, and, if put together properly was relatively reliable as a race bike.

I built a 350 from new parts, bare frame (resprayed silver) Fontana front and rear brakes, works fairing, new dry-clutch, 1967 motor, Ceriani 35mm forks etc. in Winter 1966 and Steve Jolly campaigned it during the 1967 season. I think he won about 36 races. It never broke down and we only saw the bike about every other month for routine maintenance, mainly cams and followers which were weak points!

I built the bike from a bare frame because Syd told me that just because a bike has been built by a factory, doesn't mean that everything is correct. So every aspect of the build was carefully and correctly considered, for instance, just making sure that the front wheel was central to the forks often required shims. Similarly, with the rear wheel, we used to tape a piece of string around the steering column then pull it back and line it up with the rear tyre (no seat or tank fitted) and you'd be surprised how far it deviated from the centre of the main frame tube, so we'd have to make up new alloy spacers.

Little things like that, when all put together meant that these bikes would be better and easier to ride, hence faster lap times, and more reliable (as no electrical wires broke) ensuring better results over a whole season.

That was probably the greatest thing I learned from my dad, common sense, meticulous preparation. Some people would call my dad a great tuner or a genius, but to me, watching him, a lot of the stuff he did was just good common sense. It's just that other people hadn't thought to do what he did.

A typical example is how he prepared a gear box. Most people just fitted it, as it came from the factory, but he would assemble the crankcases without the crank assembly in place, look down the hole and go through the gears checking which ones were and were not aligned correctly, then reshim them, put the crank cases together again and check the whole process all over again, readjusting the shims for a third time if required. Once again, meticulous preparation, but not rocket science, just good common sense.

When I was racing for my dad in the early 60s we didn't have that good a relationship, like quite a few father/son relationships. But looking back at it all now, I take great pride in knowing that my dad was a great racer and enthusiast, a good supporter of motorcycle racing over a long period and a guy that would talk to anybody. Ok, so he could talk the hind legs off a donkey, but I have seen many people enthralled by his conversations. Perhaps his problem was that he took on more work than he could handle, I remember Alan Cathcart waiting 3 to 4 years for his Linto engine to be done.

As for me, from the age of twelve, it was bikes, riding or working, learning or racing!

Barry Lawton.

Isle of Wight classic racer, Mike Attrill has close associations with the Lawton equipé. Here with his 350 Ala d'Oro. CRMC, Mallory Park, May 2005.

Bill Swallow

Bill Swallow (Linton Aermacchi) at full tilt, Creg ny Baa, 1994 Junior Manx Grand Prix.

I had been racing for ten years before I encountered the Aermacchi marque from a rider's viewpoint. Those formative years were exclusively astride my own Velocette-based specials, a labour of love that still goes on, but really ended competitively in 1988. By 1979 I was on my 5th Manx Grand Prix and the estimable 'Kennings' Classic Bike Championship, forerunner of the present series had just begun. My racing with the Velo had produced some great scraps with the predominant Seeleys ridden by Richard Roger, Richard Cutts, and in particular Malcolm Wheeler.

Malcolm also rode Cleve Brightman's Aermacchi and possibly on the strength of one marvellous race in 1976 when we both smashed the Cadwell lap record and finished inches apart, he recommended me for a ride on it in 1979 Manx. The records show that I came 39th, my fastest lap was just over 95mph, about the same as I was doing on the 500 Velo, and quite respectable at the time for a long stroke. The first lap was a bit of a trial, the battery spade connector kept jumping off so everything went quiet and I had to fumble about to find it and refit it without going through a hedge.

When it finally jumped off altogether, pulling up the hill from the Gooseneck it almost lost me the finish. With the bike slowing rapidly on the incline, I was desperately trying to locate the clip whilst kicking down through the gears. I did it with the bike at walking pace, as there would have been no chance of bump starting it up the hill and it was far too risky (and illegal) to go back down. So, I held it on with one hand to the pits, where a pair of pliers sorted it for the next 5 laps. Trevor White, Cleve's mate had come across to help and we struck up a partnership that lasted until his untimely death last year. I rode the Junior Classic Manx on his K4 Honda in 1986 then they banned them, so, no ride in 1987 then Trevor's illness meant that he couldn't finish the 350 Ducati he was preparing for the 1988 race. I went looking for a bike.

Cleve said I could borrow a rolling chassis but didn't have a spare Aermacchi motor for it. Jack Agnew, from Northern Ireland said he had a motor that we could use, so sponsor and friend John Poyner offered his time and his Porsche and we drove over to Temple Patrick. Jack admitted that he didn't know much about the motor but we were welcome to use it.

Sadly it turned out to be in poor condition, full of shot-blast grit and with a number of cracks in the crankcase necessitating welding, machining and resleeving of several bearing housings.

A local company who had given me some help with the Velo took on the job. I spent many hours on the phone to Syd Lawton for advice and parts. With only weeks to go I was miles off, and Syd said, 'You're not going to finish it in time, and anyway it won't be right. I do have an engine for this chap in Bristol, Bob Kent. Give him a ring.'

Bob was ever so helpful and John and I drove down and came away with his beloved long-stroke which had previously been ridden by Russell Benney (now boss of the successful 'Phase One' endurance racing team) and Dave Rumbold, and it had a reputation for being fast and reliable, and so it proved.

I finished 3rd in the Junior Classic behind brother Richard (unbeatable on Syd Lawton's Aermacchi) and Mark Linton on his dad's long-stroke.

Bob Kent's Aermacchi ran faultlessly, the only problem being the red suspension units. I mention red because it was the only good thing about them. The bike wallowed and bucked for four laps as I had a scrap on the roads with similarly mounted Alan Jackson, who I think eventually let me go for the sake of his own safety. We dropped them in the Irish Sea on the way home, but Bob was delighted and suggested I keep the bike, look after it myself and use it whenever I wanted to, my first proper racing bike. I won the pre TT classic and Billown on it, at its first visit, with a lap and race record in 1989 and did it again in the wet in 1990. John and I travelled the continent and Island with the Macchi in the back of either my Astra or his Volvo estate, (front wheel out and rear suspension strapped down). We had some great times and good results, with wins and fastest laps at Assen, Paul Ricard, Cookstown, 3rd overall in the 1,000cc class and Dundrod.

We broke the lap record at Assen with a broken swinging arm and similarly, after a hectic last lap race with Mervyn Elwood's 7R at Dundrod in 1996, returned to the pits to find that both arms were fractured and being held together basically by the torque arm and the chain! One more lap doesn't bear thinking about! I've still got that bike.

Bob's business took a turn for the worse and John and I bought the long-stoke. I'll not sell it and always said that Bob could have it back when he wanted.

1988 was my first Aermacchi ride in the Classic Manx Grand Prix and my last on the Velocette. The Velo was a bit special that year, but the opposition was getting faster. Riding on the limit I'd overshot Ballacraine twice, slid off at Ramsay Hairpin and pressed on with a broken screen and a motor burning oil, down to a broken gudgeon pin, only to run out of petrol at Kate's. Pushing in to finish 30th I was sitting disconsolately on the pit wall when Fred Walmsley offered me his Seeley for 1989. Fred didn't have a 350 and Dick Linton asked if he minded him approaching me with an offer of his own Aermacchi. So, from nothing to doubly sponsored overnight!

Pole position at Assen in 1994 with the Bob Kent Aermacchi. The piston holed on warm-up lap, so, no race for Bill.

I won on the Seeley in 1989 and was 2nd to Richard on the Linton Aermacchi in 1990, which was the start of a long and fruitful relationship with Dick.

Second again to Richard in 1991 we had a marvellous scrap in practice. Syd Lawton's short-stroke was undoubtedly quicker but Dick's original long-stroke was a lovely bike to ride.

Richard passed me down the Glencrutchery Road as we both started our second lap, and I tucked in behind him. Passing and re-passing each other, we both lapped at over 99mph. I surpassed that later, on a balmy, calm and serene evening practice, on my own for most of the session, we did 99.9mph, it just flowed and felt easy and we reckon it is still the fastest (although unofficial) lap of the course by a long-stroke. Sadly Richard died, we took on his mantle with a wet win in 1992.

Breakdowns in 1993, '94 and '95 led on to our race and lap record-smashing 1996 Junior. I'm sure, with the way both the bike and rider were going that year, the final lap of that race would have put the lap record at over 103mph and unbeaten today. But over the Mountain the bike was dragging, head-wind I thought, but in the winners enclosure the true reason was revealed, eleven (I think) broken spokes in the rear wheel and the rim pulled across so far that the tyre was binding against the swinging arm. We couldn't even push it!

For the next 4 years we pitted a 250 Aermacchi against the 2-stroke hordes. It was great fun trying and we led the race on a number of occasions before the fragility of the motor let it down. I'm sure that a 100mph lap would have been possible and we are still the fastest ever single cylinder 250 on the course at 98.5mph. I don't see that being beaten.

When Richard died, Syd offered me his own 250 ex-Angelo Bergamonti, fast but with a vibration that set your feet on fire, and able to rev to 12,000rpm, it was a revelation. I rode it in the TT Classic Lightweight at Castletown in 1993. The weather was awful and the opposition included Bob Jackson and brother Alan on 50+bhp Suzuki twins.

I took the lead and was cleaning up when I arrived at Cross-4-Ways on the penultimate lap. The front wheel locked, it still had Ferodo AM4 'Green' linings. I stayed on but had to make a three point turn up the Ballasalla Road, at which point the Jackson two came past. I caught up and relegated Alan to 3rd, but Bob was too good! So near!

I rode the bike again at Thruxton. The late Colin Breeze was out on his 350. The 250 could stay with him by slipstreaming and we swapped the lead several times. The grunt of the bigger engine took Colin past me out of the last chicane, 2nd again, but a notable event as the whole Lawton family had made a party of it. I don't think Syd went racing again.

In between riding for Dick, I went abroad and to the Isle of Man with John Oldfield and his team. Primarily racing his triples, he also had a very endearing short-stroke Aermacchi. I was sitting in the warm-up area at the 1999 Ulster Grand Prix, on the Oldfield Aermacchi when it suddenly stopped, just as the gate opened and the rest of the field streamed off for the sighting lap. A 6-volt battery and coil ignition, it had to be the battery, I surmised. The nearest van was Denis Gallagher's, he had a 12-volt battery but that was twice as big as the space available. We tied it on with bungees, cable ties and tape by which time the Clerk of the Course wouldn't let me out. I argued for the sake of the experience and he allowed me a back row grid position without the chance of the sighting lap. I caught Colin Rogers, (Honda twin) on the last lap and held it to the flag. Our lap record of 99.828mph stands today for 350s. So near and yet so far, from that 100mph lap, as there is no more classic racing at Dundrod.

There are so many wonderful racing memories and Aermacchi gave me the best days of my career. I still have an Aermacchi and I'm just going out to the garage now, to enjoy another look!

Bill Swallow.

Bill leads Derek Whalley, Ballabeg Hairpin, De Montfort pre TT Classic, 2003.

Interviewed by Murray Walker, 2002, after winning the, Junior Classic Manx Grand Prix.

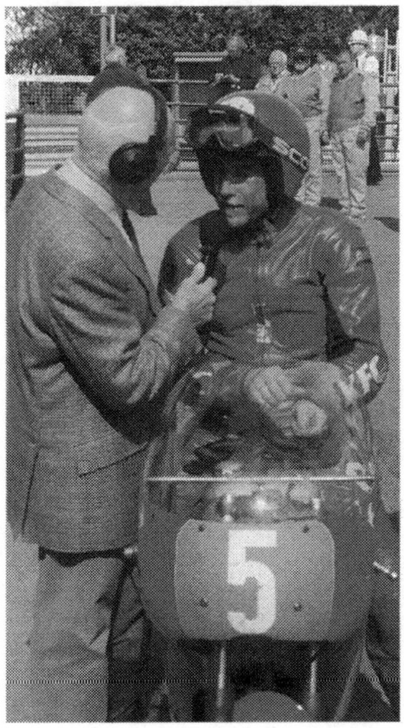

Richard's Famous Five: Ian Lawton

Richard Swallow celebrates the first 100mph lap of the TT circuit by a 350 single , in 1988, flanked by Syd and Ian Lawton.

When he won the Junior Classic in 1991, Richard Swallow became the first rider ever to win the same Manx Grand Prix race for 5 years in succession. All of these victories were achieved on my father Syd's 350cc Aermacchi. For good measure he also set the first ever over l00mph lap by a 350 single, at 100.97 - finally beating the previous best of 99.99mph set by Alan Barnett when finishing second to Giacomo Agostini in the Junior TT in 1970, also on one of Syd's bikes.

This extraordinary series of victories says a lot for Syd's preparation, but I have praised that highly elsewhere. Here I would like to concentrate on the rider. Richard was a wonderful man. Skinny as a rake, and balding on top, he was never overly bothered about his appearance. But despite his Yorkshire heritage, there was nothing 'dour' about him. He was always game for a laugh, and a joke of some kind was never far away. The joking stopped, however, when he pulled on his baggy old black leathers and plain red helmet adorned with the white rose of his native county, and came to the start line.

Richard and I raced against each other only on a very few occasions, because normally, if we were racing at the same meeting one would be on the 350 and the other on the 500. Normally I don't like to give a competitor too much respect, at least not when I intend to do everything in my power to beat him. But Richard was one of the very few riders that I admired immensely for his natural talent, not just on the roads but on short circuits as well. He just did not seem to know what it meant to have an 'off day'.

He was, basically, blindingly fast. With his slender frame he could tuck himself so far inside the fairing of an Aermacchi that absolutely nothing stuck out into the wind until he reached a corner, where without sitting up or sliding off the seat, he would stick out a gangly inner leg and plant his knee firmly on the tarmac. Coupled with his light weight, Richard was perfectly suited to the job of trying to eke the maximum out of a low-powered classic bike around the daunting TT circuit.

Syd was first introduced to Richard by mutual friend George Ridgeon, at a North Gloucester club meeting in early 1987. Syd was on the lookout for a rider for the Classic Manx races introduced in 1983, as a way of rekindling his relationship with a circuit on which he had already had great success both as a rider and tuner-sponsor. George recommended Richard, and the two hit it off immediately. A test day was arranged, Richard's pace was evident, and the rest is history.

Syd was fond of saying that he and Richard never had a cross word. For the most part that was absolutely true and we all know that under the pressures of racing that is no mean feat. But during one particular Senior Classic Manx, with Richard riding the 444, he was clearly losing time and came into the pits for an unscheduled stop. He had problems changing gear, so I checked the gear lever and something was jamming. It was internal, and we were not going to be able to fix it. But Syd, caught up in the moment as he always was, was yelling, 'Get out there and ride the bloody thing!' Richard, quite rightly, said a few choice words from under his helmet, but whether Syd heard them or not mattered little. By now everyone's body language told its own story.

There was another hilarious occasion when Richard was riding the same bike at the Classic TT races on the Billown circuit at Castletown. He was leading when a faulty plug put him out of the race. He was no Olympic athlete, but when he found out that this particular Lodge plug had been in active service since the seventies he launched it into the next field with the most tremendous throw. For once, Syd wisely realised that a period of diplomatic silence would be appropriate.

Despite these rare but inevitable hitches, they did have a tremendous relationship, as anyone who saw their hilarious double act at Manx awards ceremonies would attest. Syd genuinely regarded Richard as another son, and I also thought of him as a brother. The day after his fifth win he and I adjourned to the go-kart stadium at Onchan, and conducted a ferocious battle watched only by increasingly agitated staff. It was always going to end in upset, unfortunately that of my cart, literally! But when we met up all too rarely in London, racing was never a major topic of conversation, and the beer and friendship flowed easily.

When I finally made my own debut on the roads in 1989, at Billown in the Junior Classic TT, I finished in a safe second place, although a long way behind his equally talented brother, Bill. Richard had pulled out early on his own beloved but rather outpaced Greeves, and was spectating at the twisty section after the Iron Gates. He told me afterwards that, for a newcomer I was absolutely flying. I can pay him no greater respect than to say that his compliment probably meant more to me than any other I was ever paid in my racing career.

Richard's untimely death was a family tragedy and a terrible loss to us all. But his racing achievements, and even more his status as a man you would always be proud to call your friend, live on.

Ian Lawton.

Riding for the Old Man: Ian Lawton

Barry and Ian Lawton take a one two at Snetterton during the mid-1980s, both on 350 Aermacchis.

Aermacchis hold wonderful memories for me. I cut my teeth on them, and I was lucky enough to ride bikes prepared by my father, Syd, who you will have read about elsewhere in this book. With their light weight, and low centre of gravity, thanks to the horizontal engine layout, these bikes could always out-handle and out-brake rivals. When they were prepared by someone with the dedication, commitment and meticulous attention to detail that Syd had, they were truly sensational. Not for nothing do my contemporaries chide me that I only won races because I had 'works' bikes.

People often remember him as an engine tuner par excellence. What they forget is that he spent hours experimenting with handling set-ups, he couldn't have produced so many race-winning machines for the Isle of Man otherwise. Different spring rates at the rear, different grades of fork oil at the front - mixing in a small amount of graphite grease into a 50:50 SAE 5/10 mixture to 'ginger it up' for added stiction - and different preloads at both ends. He would regularly check for wear in the steering head and swinging arm bearings, and it was not until he was well into his seventies that he gave up lying flat on the floor to check wheel alignment after every gearing change, handing the job to the rider instead. But woe betide anyone who attempted to dodge this responsibility and trust to luck! He paid similar attention to setting up the drum brakes, and although I was never much of a 'steady it with the rear' through a fast corner merchant, I never felt - even with our big Aermacchi - that we needed a disc set-up at the all-important front. Ferodo 'green' linings with slight relief of the leading edge were always our preferred option for real bite, although one had to show them respect in the wet.

But there is no doubt that his engines were his real forte. I fondly remember the 250 machine that we built up from scratch together, and on which I started my racing career. We took it to two Classic Lightweight Race of the Year victories in 1982 - 1983, the later with a frame that would only take a 250 engine after I had reduced its steering head angle in an argument with the armco at Snetterton. In those days we had a comfortable power and handling advantage over the main 250 Ducati opposition, and in a replay of the late sixties we even held off the two-stroke Yamaha and Suzuki competition for a while. Although they out-accelerated us massively, we could always make it up on handling, braking and reliability. A best lap of 1.25.1 round the old pre-Russells-chicane Snetterton, when mixing it with modern 125 Honda singles, was testament to the bike's ability.

Initially our 250 engine was only a relatively standard short-stroke, but for 1983 Syd installed an ex-works, ultra-short-stroke, eccentric-rocker version. The first time I rode it Dennis Trollope complained when it out-accelerated his 500 Manx - even with the penalty of my thirteen stones on board - although this was perhaps not so surprising because, unusually, Syd had not placed a strip of tape on the original Smith's rev counter. I had just jumped off a 350 Yamaha, and knew that the old 250 had revved to at least 11,500 quite happily. And this one seemed to love to rev, still producing useful power right through 12,000 and even up to 13,000 on occasion. Everyone said it sounded like a 2-stroke when braking hard and blipping the throttle to shift down through the box as smoothly and fast as possible. What Syd said when I told him what I had been revving it to will remain a secret. But the fact that a 250 pushrod single could reliably produce power up to those sort of revs was a massive and arguably unequalled tribute to both the Aermacchi designers and engineers, and to Syd.

Of course the 350s were always a pleasure to ride, again in those days having a comfortable power advantage over the Ducati, AJS and Norton opposition. Only in non-Classic club meetings did we come up against the dreaded Hondas, but even my dices with Rob Claude's hard-ridden example in North Gloucester club meetings in the early eighties were always pretty equal. One of my earliest races on a 350 was against my brother Barry in a Kennings Series race in 1983 at our

Ian Lawton, 350 Aermacchi, Classic TT 1989, second place and best newcomer.

Ian Lawton rides the 249cc Aermacchi to victory in the Classic Lightweight Race of the Year, Snetterton, 1983.

local circuit, Thruxton. He had told me to follow him through the fast and tricky Village Curve, we both got away in the lead, but after we had apexed the curve, Barry was drifting further and further towards the outside edge of the circuit. At the last minute I thought 'bugger this' and eased off, only to see him disappear off into a field. He always reminds me that he stayed on though, Not long afterwards we had a great dice in the same series at Snetterton, with him just taking the honours.

Syd and I had a number of failures in Classic Junior Races of the Year. 1985 saw a particularly hectic dice between myself, Mark Linton and Dave Dearden, all Aermacchi mounted, with these two stealing each other's tarmac to such an extent that I was content to sit at close quarters and wait for the inevitable. Unfortunately I didn't leave enough room because, when they collided on the penultimate lap, Mark went down and took me with him. Much later he gave me a tape of Alan Robinson's commentary, which was a classic in its own right. Another failure in 1982 was caused by a misfire that had plagued us all season, which turned out to derive from Syd's insistence on running an ignition switch that had been on the bike since the late sixties. But we cracked it in 1989, when I was no longer racing the Aermacchis regularly, to beat a young John Barton who was just emerging as a fine talent.

Jim Porter checks his lap time with Syd Lawton as Barry Lawton looks on.

Later in 1992, with the Race of the Year switched from Snetterton to Pembrey, we had another untraceable misfire that left me having to make up time ridiculously on the brakes and in the corners against the late, great Colin Breeze and Peter Crew, again, all Aermacchi mounted. I dived inside them both from way behind at the hairpin, thought I'd left Colin room, but he cut across me, my front wheel hit him amidships and down he went. As the bike got worse and Peter pulled away, I retired disconsolate, not looking forward to what Colin's usually delightful man-mountain of a sponsor, Alf Mountford, might do to me. To make matters worse, Pete himself pulled in a few laps later, when we were so far ahead of Jerry Summerfield's Norton that I could have won the race slowing down. That showed a lack of perseverance of which I was not proud.

Our experiments with 500s were in many ways the most frustrating. The factory had produced a 382 and then a 408, in 1982 Syd produced a 422 by increasing the bore to 82mm, which required the mouths of the crankcases to be machined out and new barrels and liners made up. We campaigned this for some years, but in the Kennings series we always had a significant power deficit against the likes of Bob Newby's 4-valve Manx, Jon Cronshaw's even then rapid BSA, and a variety of G50s and the like. Probably our best result was a 3rd at Brands in 1983, where we managed to beat Dave Hughes on the infamous Arter 'wheelbarrow' Matchless and top modern rider Gary Lingham on a G50. But we were always well down on speed at faster circuits. Again in 1983 we won the Senior Classic Race of the Year by about half a lap, but in fairness, the quality of the opposition that year was not unduly high. And in the same year we also managed to win the North Gloucester club's 4-stroke single championship, up against the super-fast but slow-starting Honda of Mick Hatherill.

In the late 1980s Syd took the bike out to a 444, this time by having a completely new crankshaft with longer 84mm stroke made up. This engine was unreliable to start with, and he had to make a number of modifications to the big-end and main bearings. It was definitely faster and, although I did not ride it that much, in Richard Swallow's hands it did great things in the Isle of Man. Probably my finest memory of Syd's engineering genius was the 6-speed Shaftleitner gearbox. He spent a whole month working on that alone, endlessly stripping and rebuilding it to get the alignments and shimming absolutely perfect. When I tested it, you only had to breath on the pedal for it to change gear.

Eventually, after Syd had more or less retired, I did dynamometer testing on this engine - something he had never used before, preferring his intuition and

Ian Lawton rides the 422cc Aermacchi to victory in the Classic Senior Race of the Year, Snetterton, 1983.

experience to determine appropriate carburation, ignition timing and exhaust length settings. We increased the power to 39.5bhp at the back wheel, although this is still very low compared to the outputs of modern Classic 500s. And in a one-off outing at the inaugural Festival of Speed at Goodwood in 1993 we managed to make fastest time of the day, a record that will last forever as an official time because of Chaz Guy's unfortunate death on the last practice run, which meant that the bikes have ostensibly only paraded rather than raced ever since.

My favourite memory has to be the first International Historic Racing Organisation meeting was held at the super-fast Monza circuit, in 1990. On the 350 long-stroke, which was still extremely good on top end but just took longer to get there, I had qualified well down on Bill Swallow on similar machine. Sitting at breakfast on the morning of the race, Syd was studying the form in the programme as he always did, and then confidently announced that, 'no one will see the way Bill goes in the 350, he's a bloody good rider'. Come the race, I was hanging on to Bill for grim death but he was gradually eking out a lead that was going to be difficult for me to claw back in the few laps remaining. Then he clipped the curve in the double chicane at the start of the last lap, had to ease off momentarily, and that was my chance. I caught him through Curva Grande, almost touching his back wheel as we tucked right down inside the fairings at full throttle and lean, pulled out of his slipstream to outbreak him into the next chicane, and then pulled away to win by some distance. Initially, with us both on red bikes with black leathers, no one realised we had swapped positions.

I rode up the pit lane, handed the bike to a beaming Syd, and said quietly: 'Now who's a bloody good rider?' Did he mean to wind me up that morning as a clever ploy? I never asked him, and will never know. But what a man! Not only a great rider and tuner, but much more of a colleague and best friend to me than just a father. And a lot of that is thanks to his wonderful Aermacchis.

Ian Lawton

Ian on the 444 Lawton Aermacchi, Classic TT, 1991. A beautiful motorcycle, and a credit to its creator.

Chris and Paul: The Bladon Brothers

The Bladon Brothers, Chris, left and Paul, right, with Joey Dunlop's 250 Aermacchi, later converted to 350cc for 1994 Manx Grand Prix. All up weight 217lb (98.5kg).

In the early l970s my brother Paul purchased an Aermacchi 350 1966 Ala d'Oro racing bike through a contact from Francis Beart. Over the years we became good friends with Francis and we always admired his workmanship and meticulous attention to detail.

Having raced my own Seeley G50 for a number of years I (Chris), was very impressed by the handing and performance of the Aermacchi and entered it in the 1977 Junior Manx Grand Prix. After a trouble-free practice, the crankpin broke on the 3rd lap of the race at the end of the Sulby Straight and I watched the remainder of the race from the Ginger Hall Hotel with a pint of beer.

But clever design of the Aermacchi with the unit construction, dry-clutch, pushrod engine always belied its performance and it was always an easy bike to work on. The next step was to sell the Seeley and purchase my own Aermacchi. Over the years we raced our Aermacchis at many of the British short circuits in a vain attempt to keep up with the Japanese two-strokes, and it was good fun trying. The introduction of classic racing and the Classic Manx Grand Prix gave the

Paul Bladon at the bottom of Bray Hill during the 1983 Classic Manx Grand Prix.

Aermacchis a new lease of life as competitive racing machines and they became desirable again. We entered the first Classic Manx of 1983 on our Aermacchis where I finished 7th and Paul finished 33rd after carburation problems from the first lap. This started a long association with the Manx Grand Prix, I raced the Aermacchi in every Classic Manx from 1983 to 1989 with a best result of 4th place in 1984 and 1987.

Ian Young (Bladon Aermacchi) at the 1989 Classic Manx Grand Prix, in which he finished 2nd, lapping at over 98mph.

Over the years the constant development on our own Aermacchis formed the core of our business on engine and chassis preparation for other Aermacchi owners. Even now, years later when our main business is engine preparation on modern Japanese four-strokes we are still involved with Aermacchi motorcycles and are still preparing bikes for original owners from all those years ago.

Our most interesting involvement with Aermacchi was a request by Terry Teece from the Isle of Man to built a lightweight 250 Aermacchi for the 1993 Classic Manx Grand Prix. From an original 250cc 1966 bike supplied by Terry we built a bike with a total weight of 217lb (98.5kg) We manufactured over ninety different components for the bike including titanium wheel spindles and engine bolts, aluminium steering column, aluminium spoke nipples and many other lightweight components. Joey really liked the bike and thought the handling was excellent, but with only 25bhp at the rear wheel it was somewhat less than the usual 150+bhp Honda he was used to riding although he said he could relate it more to his 125 Honda.

The race started well with Joey lapping at over 94mph and lying in 2nd place. Coming down Bray Hill flat out in top gear he hit a bump, lifting the rear wheel off the ground, over-revving the bike and dropping a valve.

When we asked Joey what had happened he said he was enjoying it so much he got a 'wee bit carried away' and the rev-counter was coming back past 11,000rpm when it went. Lifting the bike into the van we heard a rattle in the exhaust. A moment later the head of the inlet valve fell out of the exhaust pipe. We knew then it was terminal.

For the 1994 Classic Manx Grand Prix we built Joey a less frail 350 Aermacchi using the same chassis. He finished 2nd behind Bob Heath at an average of 97.5mph.

Ian Young was a rider we sponsored on Paul's Aermacchi and he finished second in the 1989 Classic Manx Grand Prix lapping at over 98mph. Sadly, he lost his life at the TT a few years later.

Through our involvement with Aermacchi, we have met many interesting people and have made many friends over the years, and we still admire these well-designed, stylish Italian motorcycles.

Chris and Paul Bladon.

Opposite, Joey Dunlop at Tower Bends during the 1993 Classic Manx, enjoying the lightweight 250 Terry Teece machine.

Aermacchi by Mick Walker

Foale Aermacchi

The interesting Tony Foale Aermacchi, was originally built in 1971/72, It was uprated to monoshock rear suspension in 1975. When first built it featured a one-off 350 water-cooled engine with double overhead camshaft. The machine was rebuilt during the early 1980s by Dick Linton, and fitted with a 1972 249cc engine from the Varese works.

Sold to Dennis Brew in the Isle of Man during the mid-1970s, it was later owned by a private collector in the West Country. Subsequently it was purchased by Isle of Wight dealer and Aermacchi racer, Mike Attrill in 2004.

CHRIS BLADON 91

INDEX *figures in italics refer to captions*

Made in the USA
Monee, IL
13 September 2026

62980998R10118